THE ISLAMIC LAWS OF ANIMAL SLAUGHTER

In the name of Allāh,
Most Gracious, Most Merciful.
All praise be to Allāh,
Lord of the Worlds,
and peace and blessings be
upon His Messenger Muḥammad,
Mercy to the Worlds

THE ISLAMIC LAWS OF ANIMAL SLAUGHTER

A Discussion on the Islamic Laws for Slaughtering Animals and a Survey of Modern-Day Slaughtering Methods

Written in Arabic by Mufti Muhammad Taqi Usmani
Translated into English by Amir A. Toft

WHITE THREAD
SANTA BARBARA • CALIFORNIA • USA

First Edition November 2006

ISBN-13: 978-1-933764-01-6
ISBN-10: 1933764-01-5

Published by
White Thread Press
California USA
www.whitethreadpress.com
info@whitethreadpress.com
Tel: 1 805 968 4666

Distributed in the UK by
Azhar Academy Ltd. London
sales@azharacademy.com

Library of Congress Cataloging-in-Publication Data

'Usmani, Muhammad Taqi.
The Islamic laws of animal slaughter : a discussion on the Islamic laws for slaughtering animals & a survey of modern-day slaughtering methods / written in Arabic by Mufti Muhammad Taqi Usmani ; translated into English by Amir A. Toft. —1st ed.
p. cm.
Includes bibliographical references and index.
ISBN-13: 978-1-933764-01-6 (softcover : alk. paper)
ISBN-10: 1-933764-01-5 (softcover : alk. paper)
1. Slaughtering and slaughter-houses (Islamic law). 2. Animal welfare–Religious aspects–Islam. 3. Islam–Customs and practices. I. Toft, Amir A., 1984- II. Title.
KBP3123.3.U86 2006
297.5'76–dc22
2006011470

♾ Printed and bound in the United States of America on acid-free paper. The paper used in this book meets the minimum requirement of ANSI/NISO Z39.48-1992 (R 1997) (Permanence of Paper). The binding material has been chosen for strength and durability.

Cover image from iStockphoto
Cover design by CWDM
Book design and typesetting by ARM

"O people, eat from whatever is
upon the earth that is lawful and wholesome"

Qur'ān 2:168

Transliteration Key

ء (أ إ) ' (A slight catch in the breath. It is also used to indicate where the *hamza* has been dropped from the beginning of a word.)

ا a, ā

ب b

ت t

ث th (Should be pronounced as the *th* in *thin* or *thirst.*)

ج j

ح ḥ (Tensely breathed *h* sound.)

خ kh (Pronounced like the *ch* in Scottish *loch* with the mouth hollowed to produce a full sound.)

د d

ذ dh (Should be pronounced as the *th* in *this* or *that.*)

ر r

ز z

س s

ش sh

ص ṣ (A heavy *s* pronounced far back in the mouth with the mouth hollowed to produce a full sound.)

ض ḍ (A heavy *d/dh* pronounced far back in the mouth with the mouth hollowed to produce a full sound.)

ط ṭ (A heavy *t* pronounced far back in the mouth with the mouth hollowed to produce a full sound.)

ظ ẓ (A heavy *dh* pronounced far back in the mouth with the mouth hollowed to produce a full sound.)

ع ʿ, ʿa, ʿi, ʿu (Pronounced from the throat.)

غ gh (Pronounced like a throaty French *r* with the mouth hollowed to produce a full sound.)

ف f

ق q (A guttural *q* sound with the mouth hollowed to produce a full sound.)

ك k

ل l

م m

ن n

و w, ū, u.

ه h

ي y, ī, i

The following calligraphic symbols are used throughout this book as honorific expressions of endearment and reverence for those mentioned in the parentheses below:

ﷺ (Prophet Muḥammad) "May Allāh bless him and give him peace"

عليه السلام (Other prophets) "Peace be upon him"

رضي الله عنهم, رضي الله عنه (Companions of the Prophet ﷺ) "May Allāh be pleased with him, them (her)"

Contents

Translator's Foreword

THE DIET of a Muslim is unquestionably one of the most important matters that Islamic law addresses. The stern, unequivocal commands of Allāh and His Messenger ﷺ to eat what is lawful (*ḥalāl*) show why this is so. In one verse of the Qur'ān, Allāh says, "O people, eat from whatever is upon the earth that is lawful and wholesome" (Qur'ān 2:168).

In a rigorously authenticated (*ṣaḥīḥ*) ḥadīth, Abū Hurayra relates, "The Messenger of Allāh ﷺ said, 'O people! Indeed Allāh is pure; He does not accept anything but what is pure. And Allāh has commanded of the believers what he has commanded of the messengers. He says, "O messengers, eat of what is wholesome and do righteous deeds. Indeed, I am all-knowing of what you do" [(Qur'ān 23:51)]. And He says, "O you who believe, eat of the wholesome things that We have provided you"' [(Qur'ān 2:172)]. He then related the story of a person who has traveled a long way [for some righteous purpose], and whose clothes have become tattered and dirtied. He extends his hands to the heavens, saying, 'O my Lord! O my Lord!' But what he has eaten is unlawful (*ḥarām*), and what he has drunk is unlawful, and what he wears is unlawful, and he has been nourished with what is unlawful—so how can he be answered?" (*Ṣaḥīḥ Muslim*)

Furthermore, consider the well-known English saying "You are what you eat." There is a great deal of truth to this statement. Our bodies are physically composed of what we consume. Thus, by putting in them any food or drink that Allāh, the Creator of the Worlds, has deemed unlawful and impure, our bodies become reservoirs for impurity. Just as a Muslim's prayer is unacceptable if he or she is in a state of outward physical impurity, how can our prayers or our supplications or any of our other acts of worship deserve to be accepted if we offer them while impurity runs through the veins of our bodies?

Therefore, from both a practical and, more importantly, from a textual perspective, there is no question that we as Muslims have a solemn religious duty to learn the laws of Islam regarding lawful food and drink and make every effort to adhere to them. After all, it is Allāh who has created us and continues to provide us with all that we have to eat and drink. Sincere gratitude for this tremendous, uninterrupted blessing demands that we only eat what He has deemed lawful for us.

Pious People's Care for What They Ate

Many people today, Muslim and otherwise, are astonished by the tremendous achievements of past Muslim scholars, in spite of the materially and technologically inferior resources they possessed. Imām Jalāl al-Dīn al-Suyūṭī (d. 1505 CE) wrote over 1,200 independent works, many of which comprised multiple volumes. He wrote half of *Tafsīr al-Jalālayn* (the other half was written by his teacher Jalāl al-Dīn al-Maḥallī), a masterfully written and globally accepted commentary of the Qur'ān, in just forty days and did not even need to take time to revise it. ʿAllāma ʿAlī ibn ʿAqīl al-Baghdādī (sixth century AH) wrote *Kitāb al-Funūn,* a compendium on various Islamic disciplines comprising an extraordinary 800 volumes. Ibn Jarīr al-Ṭabarī, a great historian, commentator of Qur'ān, and ḥadīth scholar, was such an astoundingly prolific writer that when his students counted the pages he wrote and divided them over the days he lived from puberty until his death at 86, they found that he averaged fourteen pages a day, which would make the total almost beyond the realm of human possibility.

The accomplishments of the pious people in recent generations are no less remarkable. Mawlānā Ashraf ʿAlī Thānawī (d. 1943 CE) authored over 1,000 works, personally wrote responses to thousands of letters he received from people, and also found time to teach and sit with people to advise them and answer their religious inquiries. Shaykh al-Ḥadīth Muḥammad Zakariyyā Kāndhlawī (d. 1982 CE) has been praised by scholars from all parts of the Muslim world for his brilliance and deep knowledge of many Islamic sciences, especially Ḥadīth. He taught in the Indian subcontinent and in Madīna, and he was a prolific writer in both Arabic and Urdu. It's almost inconceivable that he wrote his first notable work—a three-volume Urdu commentary on *Alfiyya Ibn Mālik* (a classic Arabic grammar text set in poetic meter)—when he was only thirteen years old.

No doubt, Allāh endowed these luminaries with remarkable talents, but rarely does Allāh grant such phenomenal ability and long-standing honor to anyone but his obedient servants. So what was it on their part that delivered them to such greatness?

A careful examination of these figures will find that on top of their strict adherence to the Qur'ān and the Sunna, their diligence and devotion in their prayers and other acts of worship, and their undying efforts to avoid committing sin, one of the common characteristics that they possessed was tremendous particularity about what they ate. Their great care in choosing what they ate was perhaps one of the main factors in the success they had in serving Islam, and should be a lesson for every Muslim who strives to come closer to Allāh Almighty.

This Book

This book takes up the issue of slaughtered meat, which has become probably the most controversial food-related topic among Muslims. What are the precise rules for lawfully slaughtering animals? When is slaughtered meat prohibited for Muslims to eat? What exactly is Islamic law's understanding of the Qur'ānic verse "And the food of the People of the Book is permitted to you" (Qur'ān 5:5)? Even after knowing Islam's laws for animal slaughter, how do we know whether the meat commonly sold in supermarkets and restaurants in the West is lawful, since we do not know how the animals have been slaughtered? These are some of the questions on the mind of many Muslims, and this book answers them.

The author, Muftī Muhammad Taqi Usmani, is recognized around the world for his knowledge. He is one of the world's foremost experts in Islamic Law, Ḥadīth, and Economics. He is the deputy chairman of the Jeddah-based Islamic Fiqh Council of the Organization of Islamic Conference (OIC), and for the last four decades he has been teaching at Darul Uloom in Karachi, Pakistan, an institute of higher Islamic learning established by his late father, Muftī Shafīᶜ Usmani (may Allāh have mercy on him). He has authored over forty books in Arabic, Urdu, and English on various topics. This book he wrote in Arabic with the title *Aḥkām al-Dhabā'iḥ*.

The modern-day food industry goes far beyond slaughtered meat, and much more work in the area must be done. Even this book is, in a way, incomplete. Particularly in a couple of sections concerning meat slaughtered en masse by machine in today's slaughterhouses, the author hesitates to pass any firm legal ruling, citing a lack of knowledge of the ins and outs of modern slaughtering techniques, and invites those with expertise in such matters to conduct their own research and contribute to the cause of arriving at the correct Islamic resolution. This book, though, is the crucial first step, as there has not yet been a published work (at least not in the English language) that thoroughly examines the Islamic laws of animal slaughter from a scholarly perspective and is, at the same time, accessible to a general Muslim audience seeking clarity on the issue. It is my hope and belief that this book will, Allāh willing, create greater awareness among Muslims, especially those living in the West, about the importance of eating lawful food, and will serve as a stepping stone for other authentic Muslim scholars to work hand in hand with experienced Muslim food experts to navigate the massive informational web that is the food industry and provide Muslims at large with clear, comprehensive reports of which available foods are lawful according to the laws of Islam and which are not.

Modifications and Appendices

In bringing this work into the English language, I labored to translate the text as faithfully as possible. However, in a few places where fastidious fidelity came at the expense of readability, I found it prudent to make certain small modifications.

Though the original text does have section headings, it does not have numbered chapters or clearly demarcated subsections. To make the arrangement of the book more clear, the translated text has been divided into an introduction and three numbered chapters. The three chapters contain their own sections and subsections, which correspond to the sections and subsections of the original text. A couple of logical headings not present in the original text, however, have been added for clarity, though no content in these sections has been added, altered, or omitted whatsoever.

In the original Arabic, the author uses footnotes for source references and for additional commentary about the text. In this translation, the source references are incorporated into the body of the text. All source references are left as the author cites them in the original text and have not been adjusted to correspond to different editions from what the author used. Furthermore, the Arabic text does not contain a bibliography, and as I did not have available the editions that the author used for most of the sources, the bibliography added at the end of this translation mainly just lists the authors and titles of these sources. For the few works whose editions I had matched those of the author, I have provided the full bibliographical information.

As for the author's own comments, they have been translated and placed in footnotes. Additionally, there were several translator's comments that were deemed necessary to clarify a few parts that might be confusing in translation, especially to one uninitiated in the advanced Islamic sciences of Ḥadīth and Jurisprudence. Each of these comments is marked with "Translator" at the end of the note.

This translation also includes a detailed glossary of important Arabic terms used, as well as a subject index.

The concluding section of the original text is a survey on meats imported into Muslim countries. In it the author reproduces verbatim a number of reports written by Muslim officials from Saudi Arabia who traveled to non-Muslim countries that exported slaughtered meat to Muslim countries. In short, their reports reveal that the animals whose meat is exported to Muslim countries are not slaughtered in accordance with the prescribed Islamic method and are thus not lawful. The complete translation of this survey has been deemed unnecessary and thus has not been included in this book. Instead, a summary of the reports may be found in Appendix A, and any scholar or researcher interested in reading the full reports may refer to the original text. Following this, Appendix B contains the author's summary of the book.

The remaining two appendices have been added to the book, and the reader should take care to read these after completing the main text. These articles have been appended because they discuss matters of great relevance to the Western Muslim audience, especially those in the US. The author's work itself does not, and probably should not, touch on these specific issues, as its main objective is to reach out to a broad swathe of Muslim societies by clearly outlining the basic rules of the animal slaughter and dispelling misunderstandings that are generally widespread the world over, not to delve into questions whose significance is limited to a Western audience. So the dominantly American and British readership would do well not to overlook these short addenda.

Appendix C, written by Muftī Abdur-Rahman ibn Yusuf, contains a brief essay on the so-called *zabiha*-meat issue—a topic of special importance for Muslims in the West, especially in the United States.

On the one hand, there are Muslims who insist that simply reciting the name of Allāh over slaughtered meat before eating it renders it *ḥalāl*, whereas others maintain that it is necessary for meat to be what has come (mainly in the US) to be called *zabiha*—namely, from an animal that has been hand slaughtered correctly according to Islamic rites. This issue has left many Muslims confused as to what the words *ḥalāl* and *zabiha* actually mean. The confusion is partly due to misunderstanding of the rules of animal slaughter, and partly due to misusage of the words *ḥalāl* and *zabiha*. Muftī Abdur-Rahman ibn Yusuf clarifies this issue in light of the information presented in the book.

Finally, Appendix D contains the text of "Is Islamic Slaughtering Cruel to Animals?" an article by Dr. Aisha El-Awady, an Egyptian parasitologist. In it, Dr. El-Awady compares the Islamic and Western ways of slaughtering animals and concludes that the Islamic method is the most humane for the slaughtered animal and the most healthful for the consumer.

Acknowledgments and Final Remarks

Above all, I thank Allāh for giving me the ability, health, and energy to prepare this translation. Second, I thank my parents. Without their love, support, and prayers for me throughout my life, I am doubtful whether I would now be able to present this work. I am eternally indebted to them, and my debt but increases with any work that I do.

There is one other man whose love and friendship to me and my family has served as a foundation for my pursuit of Islamic knowledge: Shaykh Ahmad Zaki Hammad. Although he had no hand in any of the actual stages of producing this work, his is the hand that sowed the seed of my love of Islam and desire to serve it, and I pray to Allāh that, if this product of mine renders any meritable service to Religion of Truth, He grant this man and his family the full share of reward that they deserve.

Before I set pen to paper, I sought the help of a number of *ʿulamāʾ*, many of whom are my own teachers, to ensure that I correctly understood the jurisprudential concepts presented in the original text. I extend particular thanks to Mawlānā Ubaidullah Saleem, my teacher for many years, who took many hours out of his impossibly busy schedule to review most of the book with me and help me to understand some of the more intricate concepts. I also thank Mawlānā Abdullah Saleem, Muftī Aziz Ahmad, and Mawlānā Sulaiman Saleem for the invaluable answers they gave to questions that puzzled me. My gratitude also goes out to my teacher Muftī Yasir Nadeem and to Orhan Hodja and Talha Alp Hodja, who all made what they would probably consider to be miniscule contributions but in fact helped me solve with confidence several passages whose translations long bedeviled me and indeed stalled the publication of this work for what seemed to me like an eternity. I would therefore be remiss not to give them some acknowledgment, for with their aid, I was able to at last crest the final torturous hill of revision.

For general copyediting, I thank, above all, my brother (who, at his request, will remain unnamed). His thorough review caught a number of translation errors and inconsistencies, and his intelligent comments throughout made the language more smooth and cohesive. My father, himself a meticulous wordsmith, also reviewed the entire translation with a close eye for clear and correct language usage; his comments made the English much more precise and cleared up what would have been confusing ambiguities.

Muftī Abdur-Rahman ibn Yusuf reviewed the entire manuscript and provided important corrections and suggestions. Furthermore, he and the team at White Thread Press have done their usual impeccable job at all stages of production in publishing this book. May Allāh continue to make the work of this blessed organization successful.

I also send my appreciation to anyone else who may have helped in any way. The absence of their names here in no way detracts from the value of their assistance. Their reward is with Allāh, and His reward surpasses any meager thanks that I may give: "Indeed, what is with Allāh is better for you, if only you were to know" (Qur'ān 16:95).

I have done my very best to render this book into something reminiscent of flavorful English while still maintaining the full import of the original Arabic—a truly daunting task, I now know. But despite my best efforts, even with a legion of helpers, any endeavor of mine is inherently flawed. I pray, and I ask that the reader pray, that Allāh forgive me, my family, my teachers, and all who helped make this translation possible. And I ask Allāh to overlook my shortcomings and accept this imperfect effort of mine.

And the last of our prayers is, "Praise be to Allāh, Lord of the Worlds."

Amir A. Toft
March 22, 2006 CE | Ṣafar 8, 1427 AH

Introduction

ALL PRAISE IS for Allāh, Lord of the Worlds. And may peace and blessings be upon the Noble Messenger Muḥammad, upon his entire family, and upon his Companions and the righteous ones who follow in their footsteps until the Day of Judgment.

Allāh Almighty has made it lawful for Muslims to eat the meat of pure animals and to take benefit from their various parts. But He has made this lawfulness contingent on adherence to firm injunctions presented in the Qur'ān and the Sunna. The driving principle behind these injunctions is that an animal, in essence, bears similarity to a human being in that it has a soul and perception and can sense comfort and pain. Looking from this perspective, the seemingly logical ruling is that it would not be lawful for a human being to slaughter an animal, eat its meat, and take benefit from its various parts. However, Allāh has made man the most noble of creation and has placed the world at his disposal. All that He has created is for man's advantage. He says, "It is He who created for you all that is in the earth" (Qur'ān 2:29).

It is clear that Allāh has made the consumption of animals permissible purely out of His vast kindness. As this is the case, He has made consuming them subject to some "ritual" (*taʿabbudī*) laws [those that are strictly and precisely determined by Allāh through the Qur'ān and the Sunna of His Messenger ﷺ]. A servant of Allāh who abides by these ritual laws thereby displays his acknowledgment that the lawfulness of an animal is a great blessing from Allāh, and that he really does not deserve to derive benefit from or take relish in eating from an animal until he recognizes this blessing, shows thanks for it, and follows the method that Allāh has prescribed for slaughtering the animal.

Of all law systems, Islamic law (Sharīʿa) is exceptional in its treatment of the issue of animal slaughter. It delineates in superb detail the acceptable slaughtering methods, founding them on sound principles from the Qur'ān and Sunna. In view of the thorough attention Islamic law pays to this matter, the act of slaughtering an animal is not a "non-ritual" (*ʿādī*) act. That is, it is not an act for which one need not adhere to any principles or laws; one may not do it however he pleases, according to his need, personal interest, or simply what he finds easiest. It is, rather, a ritual act, for which it is obligatory to conform to the laws outlined in the Qur'ān and Sunna.

The conclusion arrived at by Muftī Muḥammad ʿAbduh and his student Shaykh Rashīd Riḍā—that the slaughtering of animals is a non-ritual act that one may freely do however he wishes[1]—is an egregious error and a statement in utter conflict with clear Islamic texts. The Messenger of Allāh ﷺ states in a rigorously authenticated (*ṣaḥīḥ*) ḥadīth:

> Whoever prays as we [Muslims] pray and faces the Kaʿba [during prayer] and eats our slaughtered animals, that is a Muslim for whom is the protection of Allāh and His Messenger. (*Ṣaḥīḥ al-Bukhārī*)[2]

Even clearer than this is another narration related by Anas ibn Mālik ؓ, in which the Messenger ﷺ said:

> I have been commanded to fight mankind until they say, "There is none worthy of worship except Allāh (*lā ilāha illa 'Llāh*)." If they say this,

1 See Muḥammad ʿAbduh's *Tafsīr al-Manār.*

2 *Kitāb al-Dhabā'iḥ wa 'l-Ṣayd, Bāb Faḍl Istiqbāl al-Qibla.*

> pray as we [Muslims] pray, face the Kaʿba [during prayer], and slaughter [animals] as we do, their blood and wealth become sacred for us, [and neither of these may we violate] except when there is due right. (*Ṣaḥīḥ al-Bukhārī*)[3]

In this ḥadīth, the Messenger of Allāh ﷺ mentions the slaughtering of animals in the same context as the ritual prayer (*ṣalāt*) and facing the Kaʿba when praying. He places it among the distinguishing features of Islamic law that separate a Muslim from a non-Muslim. He makes it one of the symbolic Islamic rituals, which show that one is a Muslim and by which one's blood and wealth become safeguarded. Whose statement could prove more conclusively than the Prophet's ﷺ—such as these, as well as others—that slaughtering an animal through a legally prescribed method is indeed a ritual act and is a fundamental religious rite that affirms the very faith of the one who does it? Ḥāfiẓ Ibn Ḥajar al-ʿAsqalānī (may Allāh have mercy on him), in explanation of this ḥadīth, writes:

> The import of this ḥadīth is that the affairs of people are judged by their outward appearance. Therefore, if anyone outwardly fulfills the fundamental rites of Islam, all of the laws protecting Muslims come into effect for him, so long as he does not openly do anything that negates [his fulfillment of these rites]. (*Fatḥ al-Bārī* 1:197)

Were the slaughtering of an animal not subject to specific laws, then any animal killed by, say, a polytheist or atheist would be permissible, and lawfulness would not be restricted to the slaughter of a Muslim or *Kitābī* (member of the *Ahl al-Kitāb,* or People of the Book; i.e., a Jew or Christian).[4] Furthermore, Islamic law makes it clear that for foods of non-animal origin, such as those made from plants, it is not required that their producer be a Muslim or Kitābī. Consumption of these foods is permissible regardless of the religious affiliation of their producer. So

3 *Kitāb al-Dhabā'iḥ wa 'l-Ṣayd, Bāb Faḍl Istiqbāl al-Qibla.*

4 For simplicity, the transliterated Arabic term *Kitābī* will be used throughout instead of "a member of the People of the Book." (Translator)

if slaughtering were something unrestricted by ritual (*taʿabbudī*) laws, it would be permissible to eat an animal regardless of the religious affiliation of the one who slaughtered it. In the case of animal slaughter, therefore, the well-known fact that the slaughterer must be a Muslim or a Kitābī is in itself proof that foods of animal origin have a special place in Islamic law. For the permissibility to consume them, there must be complete conformity with the laws of slaughter laid down in the Qur'ān and Sunna.

It is no surprise that the laws for animal slaughter, as well as for hunting, have become one of the most important topics of Islamic jurisprudence (*fiqh*). Drawing from the Qur'ān, the Sunna, and the statements of the Companions ﷺ, the jurists have expounded these laws, such that not a single recognized book of jurisprudence lacks a chapter on "Hunting and Slaughtering."

In this book, I do not intend to examine the minute intricacies of these laws. The purpose instead is to clarify their legal basis and arrive at an understanding of their application in modern-day situations. I have divided the book into the following basic parts:

I. The Requirements of Islamic Law for Animal Slaughter
II. People of the Book: Slaughtered Animals and Related Issues
III. Modern-day Automated Methods of Animal Slaughter

We ask Allāh to give us success in arriving at what is correct and to guide us to what brings us His pleasure.

I

The Requirements of Islamic Law for Animal Slaughter

THE ARABIC WORD used in Islamic law to denote "lawful slaughter" is *dhakāt* (or, equivalently, *tadhkiya*). *Dhakāt*, at its lexical root ذ ك و, means "to attain completion or perfection" (*itmām*). From this root also comes *dhakā*, or "intelligence," which is a sort of completeness in its own sense, in that it gradually develops with age and experience (*Lisān al-ʿArab* 14:288).[5] Similarly, the prescribed method of slaughtering is called *dhakāt* because it completes—in other words, it fulfills—the requirements by which eating an animal becomes permissible. Imām Qurṭubī (may Allāh have mercy on him) explains the words of Allāh "except that which you duly slaughter (*dhakkaytum*)" (Qur'ān 5:3): "That is, you have carried out the slaughtering to completion" (*Tafsīr al-Qurṭubī* 6:51). Some scholars have said that lawful slaughter is named *dhakāt* because of another meaning of the word: "to produce a fragrant aroma." The technical term they use is *rā'iḥa dhakiyya,* meaning "fragrant aroma." When an animal's blood flows out during slaughtering, its meat takes on a fresh, fragrant scent.

5 Search under *dhakā* (ذكا).

This is the lexical meaning of *dhakāt.* Regarding its technical meaning, Imām Qurṭubī mentions "that it consists of (1) *dhabḥ,* or causing blood flow by cutting the jugular veins,[6] for a *madhbūḥ,* which is an animal slaughtered in the conventional way of slitting the throat;[7] (2) *naḥr,* or stabbing the lowermost part of the throat, for a *manḥūr,* which is any animal (e.g., a camel) that is typically killed in this manner;[8] and (3) *ʿaqr,* or a fatal wounding,[9] for an animal not in one's physical control, provided that one target the animal for the sake of Allāh and mention His name before striking the animal" (*Tafsīr al-Qurṭubī* 6:52–53). Since some of the fine details of the definition of *dhakāt* are disputed among Muslim scholars, it is best to define it as follows: "To kill an animal through the legally prescribed method that makes its meat lawful (*ḥalāl*) for a Muslim."

The legal requirements of a complete slaughter, as mentioned by the jurists, can be broken into three basic parts: (1) the method of killing the animal, (2) mentioning the name of Allāh when slaughtering, and (3) the religious status of the slaughterer. We will now discuss these in some detail, and we seek Allāh's help in doing so.

---

6 Refer to Appendix C, where you can read about the *zabiha*-meat issue and gain a better understanding of the word *dhabḥ.* (Translator)

7 *Dhabḥ* and *madhbūḥ: Dhabḥ* is the most common method of slaughtering animals, whereby the slaughterer manually slits the animals throat and causes its blood to drain out. This type of slaughtering is the main focus of this book. Any specialized methods of slaughtering animals, such as for hunting, receive little attention but are clearly distinguished when discussed. An animal slaughtered by *dhabḥ* is called *madhbūḥ.* (Translator)

8 *Naḥr* and *manḥūr:* This method is typically, though not necessarily, reserved for camels. The jugular veins are not cut in this method. An animal slaughtered by *naḥr* is called *manḥūr.* (Translator)

9 *ʿAqr:* Refers to a hunting situation. Since the hunter does not have physical control of the animal and cannot slaughter it in the conventional, manual way (*dhabḥ*), he may use a sharp hunting projectile (e.g., an arrow) or a hunting animal (e.g., a dog) to kill the target. (Translator)

Method of Killing the Animal

Causing Blood to Flow Out

The killing method that Islamic law considers satisfactory for a proper and lawful slaughter differs for different types of animals. In the case of an animal that is not under one's physical control, either because it is wild or because it has strayed from other domesticated animals, it is sufficient to wound it with any sharp tool that causes its blood to flow out until it dies. This type of slaughter is called "forced slaughter" (*dhakāt iḍṭirāriyya*): one cannot gain control of an animal to slaughter it by hand and is therefore "forced" to kill it from a distance. It is not necessary that such an animal be slaughtered by the methods of *dhabḥ* or *naḥr*. Forced slaughter is prescribed in the case of hunting. The bulk of this book, however, is not concerned with explaining the laws of this type of slaughter, but focuses instead on the type called "voluntary slaughter" (*dhakāt ikhtiyāriyya*).

Voluntary slaughter refers to when an animal is under one's physical control, be it a domesticated animal or a captured wild animal. In this case, it is obligatory to cause its blood to flow out by the conventional method of cutting into its jugular veins. The support for this is in the following four ḥadīths:

(1) Rāfiʿ ibn Khadīj ﷺ narrates in a long ḥadīth that his grandfather asked the Messenger of Allāh ﷺ, "May we slaughter using a cane?" The Messenger of Allāh ﷺ replied:

> [As long as you use] whatever causes the blood to flow out and the name of Allāh has been mentioned [before slaughtering], you may eat [the animal]. (*Ṣaḥīḥ al-Bukhārī*)[10]

The question is about *dhabḥ,* and *dhabḥ* consists of cutting into the jugular veins, as Imām Bukhārī reports from ʿAṭāʾ ibn Abī Rabāḥ (*Ṣaḥīḥ*

10 *Kitāb al-Dhabāʾiḥ wa 'l-ṣayd, Bāb al-Tasmiya ʿala 'l-Dhabīḥa.*

al-Bukhārī).[11] The question and answer in this ḥadīth together demonstrate that a lawful slaughtering occurs by cutting into the jugular veins such that the blood flows out.

(2) It is reported that Ibn ʿAbbās and Abū Hurayra ﷺ said:

> The Messenger of Allāh ﷺ prohibited us from the *sharīṭa* of Satan—that which is slaughtered, then skinned, and then left to die without its jugular veins having been slit. (*Sunan Abī Dāwūd*)[12]

Ibn al-Athīr, a ḥadīth scholar, historian, and lexicographer of the sixth and seventh centuries AH, states:

> A *sharīṭa* is a she-camel, and the like, whose neck is slightly slit—like the slit made by one who does blood cupping. Its jugular veins are not cut, nor is its blood made to completely flow out. This was the practice of the people before Islam, that is, during the Days of Ignorance. They would make a slight cut in the animal's neck, and that to them would be sufficient for slaughtering. They attributed this to Satan, as if he made them do it. (*Jāmiʿ al-Uṣūl* 4:482)

(3) ʿAdī ibn Ḥātim al-Ṭāʾī ﷺ relates:

> I asked, "O Messenger of Allāh, if one of us gets hold of some game [while hunting] and does not have a knife with him, may he slaughter using a flint-stone or the [pointed] tip of a staff?" He replied, "Make the blood flow out with whatever you wish and mention the name of Allāh Almighty [before slaughtering]." (*Sunan Abī Dāwūd*)[13]

Imām Nasāʾī (may Allāh have mercy on him) reports this same ḥadīth as follows:

> "[O Messenger of Allāh,] I sent out my hunting dog and caught game, but I did not find anything to slaughter it with. May I slaughter using

11 *Kitāb al-Dhabāʾiḥ wa 'l-ṣayd, Bāb al-Naḥr wa 'l-Dhabḥ.*

12 *Kitāb al-Aḍāḥī, Bāb al-Mubālagha fi 'l-Dhabḥ.*

13 *Kitāb al-Ḍaḥāyā, Bāb al-Dhabīḥa bi 'l-Marwa.*

a flint-stone or a staff?" He replied, "Make the blood flow out with whatever you wish and mention the name of Allāh Almighty [before slaughtering]." (*Sunan al-Nasā'ī*)[14]

(4) It is reported that ʿAbdullāh ibn ʿAbbās ﷺ said:

"[As long as you use] whatever cuts into the jugular veins, then you may eat [the animal you slaughter]." (*Muwaṭṭā' Imām Mālik*)[15]

On the basis of these ḥadīths and others like them, the jurists have stipulated that for a lawful slaughter, at least some of the jugular veins (*awdāj*) must be cut into. *Awdāj* is the plural of *wadaj,* which for our purposes simply means "jugular vein." Of the jugular veins, there are two main, external ones (called in Arabic *wadajān,* the dual form of *wadaj*). Ibn Manẓūr, the famous Arabic lexicographer of the seventh and eighth centuries AH, quotes from Ibn Sayyidihī: "The *wadajān* are two veins extending from the head and the lungs, and the plural is *awdāj*" (*Lisān al-ʿArab* 2:397).[16]

The jurists, however, have extended the usage of the word *awdāj* to include the trachea and esophagus. ʿAllāma Kāsānī states, "The *awdāj,* then, are four: the trachea, the esophagus, and the two [external jugular] veins between which lie the trachea and the esophagus" (*Badā'iʿ al-Ṣanā'iʿ* 5:41). The trachea is the air passageway, and the esophagus is the food passageway. There is no dispute that the most perfect slaughter is to cut into all four of these passageways—the trachea, esophagus, and both external jugular veins (*Al-Mughnī* 11:45). But the scholars have different statements regarding the case when some are cut but others are not.

Imām Shāfiʿī (may Allāh have mercy on him) says that it is obligatory for one to cut into the trachea and esophagus, and that this is sufficient

14 *Kitāb al-Aḍāḥī, Bāb Ibāḥat al-Dhabḥ bi 'l-ʿŪd;* the transmission chain of this ḥadīth contains Murayy ibn Qaṭarī al-Kūfī, whom Ibn Ḥibbān has included among his trustworthy narrators. ʿAllāma Dhahabī, however, says, "He is unknown (*majhūl*)" (*Tahdhīb al-Tahdhīb* 1:99).

15 *Kitāb al-Dhabā'iḥ, Bāb Mā Yajūzu min al-Dhakāt fī Ḥāl al-Ḍarūra.*

16 Search under *wadaj* (ودج).

for a complete slaughter, even if one does not cut either of the external jugular veins (*Fatḥ al-Bārī* 9:641; *Kitāb al-Umm* 2:259).

There are different opinions reported of Imām Mālik (may Allāh have mercy on him). The dominant one, though, as mentioned in the Mālikī books of jurisprudence, is that it is obligatory to cut into the trachea and both external jugular veins, but not the esophagus (*Al-Dhakhīra* 4:133).

There are also different opinions reported of Imām Aḥmad ibn Ḥanbal (may Allāh have mercy on him). One agrees with that of Imām Shāfiʿī, but another states that it is obligatory to cut into both external jugular veins along with both the trachea and esophagus. So he seems to stipulate cutting into all four passageways. (*Al-Mughnī* 11:44–45)

Imām Abū Ḥanīfa (may Allāh have mercy on him) states that if one cuts into three of the four passageways, whichever three they may be, and leaves one uncut, the animal is lawful. Imām Abū Yūsuf (may Allāh have mercy on him) states that it is not lawful until one cuts into the trachea, esophagus, and one of the two external jugular veins. And Imām Muḥammad (may Allāh have mercy on him) states that it is not lawful until one cuts into most of every one of the four passageways. (*Badāʾiʿ al-Ṣanāʾiʿ* 5:41)

Despite the jurists' varying opinions regarding these secondary details, they have agreed that for voluntary slaughter, the cutting point (*madhbaḥ*) is where the throat meets the upper part of the chest, and that it is necessary to cut into more than one of the four passageways. From what we have seen so far, it should be evident that the opinion requiring at least one of the two external jugular veins to be cut into is the dominant opinion. This is because making the blood flow out entirely, which is required, can only occur by cutting into one or both of the external jugular veins, which are the main blood vessels in the neck. ʿAllāma Qarāfī (may Allāh have mercy on him) says:

> The statement of the Prophet ﷺ [ḥadīth 1, p. 27] emphasizes that [it is necessary to cut one of the two external jugular veins], since causing blood to flow out can only occur via the jugular veins. The etymologi-

> cal root of *inhār* (to cause something to flow out) means "spaciousness." From this root also comes *nahr* (river) for its ability to contain water, as does *nahār* (daytime) for the way light spreads out during the daytime. (*Al-Dhakhīra* 4:133)

The proof for Imām Abū Ḥanīfa's opinion is that three of the passageways stands in place of all four, and so cutting just three of them is like cutting all of them. The guiding rule here is the famous principle of Islamic jurisprudence that in legal issues bearing some flexibility, the majority takes the ruling of the whole.[17] And although, as we have just seen, the jurists disagree as to precisely how an animal is to be slaughtered, they do not disagree that there is indeed flexibility in matters of lawful slaughter, and so this principle applies here. (*Badā'iᶜ al-Ṣanā'iᶜ* 5:42)

The Slaughtering Implement

The jurists have agreed that for a complete and proper slaughter, according to Islamic law, the implement used must be sharpened such that it easily cuts and pierces the animal because of its sharpness, not because of its heaviness. It is not necessary that the slaughtering tool be a knife; it is permissible to slaughter using anything with a sharp edge, whether it is made of steel, stone, wood, or something else. The proof for that is in a ḥadīth reported by Imām Bukhārī, Imām Muslim, and others, on the authority of Rāfiᶜ ibn Khadīj ﷺ, who relates:

> I said, "O Messenger of Allāh, tomorrow we will meet the enemy in battle, but we do not have any knives with us. May we then slaughter

---

17 For someone without a little background in the Islamic sciences, this tersely stated principle, even when read in Arabic, may likely prove enigmatic, even after multiple readings. A simple, non-legal example should help clear this up. If a businessman today wears a dress shirt that is predominantly white but has fine blue pinstripes, we would say, if asked to choose one color to describe the shirt, that it's white. We would have a hard time convincing anyone that the shirt is blue. So although the shirt does contain a color other than white, the shirt as a whole is, in effect, white because that is the color of the majority of its fabric. Similarly, in the juristic issue we discuss here, though in reality cutting three passageways misses one, we consider that all are in fact cut because the majority has been cut. (Translator)

> [our animals] using canes?" He replied, "[As long as you use] whatever causes the blood to flow out and the name of Allāh has been mentioned [before slaughtering], then you may eat it; but do not use teeth or nails." (*Jāmiᶜ al-Uṣūl* 4:489)

A moment ago, we saw a ḥadīth narrated by ᶜAdī ibn Ḥātim ﷺ, in which he asked the Messenger of Allāh ﷺ about slaughtering with a flint-stone or the pointed tip of a staff. The Messenger of Allāh ﷺ replied, "Make the blood flow out with whatever you wish."

Both of these ḥadīths (and all other related ones) agree that the slaughterer must make the blood flow out by using something that can cut by *piercing* the flesh. The requirement that this tool be sharpened is the verdict of the consensus (*ijmāᶜ*) of the recognized Muslim jurists. However, they disagree in the matter of teeth and nails. The Imāms of the Ḥijāz (Western Arabian Peninsula) have concluded that it is impermissible to slaughter using them, whether they are attached to or detached from the body. They base this on the general import of the ḥadīth of Rāfiᶜ ibn Khadīj, mentioned above, in which the Messenger of Allāh ﷺ excludes teeth and nails as acceptable tools for slaughtering.

On the other hand, Imām Abū Ḥanīfa (may Allāh have mercy on him) has interpreted the ḥadīth to refer to teeth and nails when they are still attached to the body, because in that case, death might occur by strangulation. His view is that when they are separated from the body, the slaughtering is acceptable, though it is disliked (*makrūh*). (*Radd al-Muḥtār* 5:208)

Killing Without Cutting the Jugular Veins

When an animal is killed without its jugular veins being cut, a lawful slaughtering does not occur if the animal is in one's physical control (i.e., voluntary slaughter). Aside from a few anomalous opinions, the jurists have unanimously agreed on this. Allāh says:

> Forbidden to you is [the eating of] carrion and blood and the flesh of swine and that over which other than [the name of] Allāh has been

> invoked and what has been strangled and what has been beaten to death and what has fallen to [its] death and what has been gored to death and what has been eaten by a predator—except that which you duly slaughter [before it dies]. (Qur'ān 5:3)

Ḥāfiẓ Ibn Kathīr (may Allāh have mercy on him), the well-known historian and Qur'ān commentator of the eight century AH, states in his commentary of this verse:

> "What has been strangled" (*munkhaniqa*)—an animal that has died by strangulation,[18] whether on purpose or by accident, such as when it gets tangled up in its tie and suffocates—is prohibited.
>
> "What has been beaten to death" (*mawqūdha*) is an animal that has been beaten with a heavy, unsharpened object until it dies. Ibn ʿAbbās and others have said, "It is an animal that has been beaten with a piece of wood until it passes out and dies." Qatāda states, "Before Islam, the people of the Days of Ignorance used to beat an animal with their staffs, and then, when it died, they would eat it." Furthermore, it is reported in a rigorously authenticated ḥadīth that ʿAdī ibn Ḥātim relates, "I said, 'O Messenger of Allāh, I threw a featherless arrow (*miʿrāḍ*)[19] at some game and struck it.' He replied, 'If you throw the featherless arrow and it pierces[20] [the animal], then you may eat it. But if it strikes with the

18 The verse states that "what has been strangled" (*munkhaniqa*) is prohibited. *Munkhaniqa* comes from the root *khanq*, which denotes the cutting off of air to a breathing creature. This includes not only strangulation, but also suffocation and anything that might otherwise asphyxiate the creature. Properly understanding this prohibited category is important in assessing modern methods of slaughtering animals, which is the topic of Chapter 3. (Translator)

19 *Miʿrāḍ:* "An arrow with neither feather nor arrowhead. It travels lengthwise and strikes with its middle part, not with its sharp tip" (*Lisān al-ʿArab* 9:42). Mentioned in *Tāj al-ʿArūs* is: "It is made from wood and is thin at its two ends and thick in the middle, like a piece of wood used to separate cotton from its seeds. When one throws it, it travels lengthwise and strikes with its middle part, not with its sharp tip. . . . And if a target comes near, it strikes it with the place of the arrowhead and wounds it. Related to this is the ḥadīth of ʿAdī ibn Ḥātim" (*Tāj al-ʿArūs* 5:50).

20 *Khazaqa* is the word used here for "it pierces." In reference to an arrow, it means that when it strikes a target, it penetrates it and causes it to bleed. See *Lisān al-ʿArab* and *Tāj al-ʿArūs.*

middle part, then it has been beaten to death, so do not eat it.'" Thus, he distinguished between an animal struck by an arrow or spear or something similar in its sharpness—this he made lawful—and one struck by a dull object—this he deemed to be beaten to death and did not make it lawful. This is the ruling agreed upon by the jurists.

"What has fallen to its death" (*mutaraddiya*) is that which falls from a mountain or an elevated place and consequently dies. Such an animal is not lawful. ʿAlī ibn Abī Ṭalḥa relates on the authority of Ibn ʿAbbās, "'What has fallen to its death' is that which falls from a mountain." Qatāda states, "It is that which stumbles into a well." Suddī states, "It is that which falls from a mountain or stumbles into a well."

"What has been gored to death" (*naṭīḥa*) is that which has died because another animal butted it with its horn. This animal is prohibited, even if the horn wounds it and even if the blood flows from [the throat, which is] the cutting point of slaughtering. . . .

"What has been eaten by a predator" (*mā akala 'l-sabuʿu*) is that which a lion, cheetah, tiger, wolf, or dog has pounced on and partially eaten, causing it to die. Such an animal is deemed unlawful by the consensus of the scholars, even if blood has flown out from the cutting point. During the Days of Ignorance, if a predator ate some part of a sheep, camel, cow, or the like, the people of those days would eat whatever the predator left behind. Allāh then prohibited this to the believers.

The words "except that which you duly slaughter [before it dies]" (*illā mā dhakkaytum*) refers back to whatever it can possibly refer back to—that is, those animals whose potentially death-causing event has taken place but are found and properly slaughtered while they are still alive. These are, namely, "what has been strangled," "what has been beaten to death," "what has fallen to its death," "what has been gored to death," and "what has been eaten by a predator."[21] Regarding the meaning of the words "except that which you duly slaughter," ʿAlī ibn

21 Out of the items prohibited in the verse, these five are the only ones that fall into this final category. The rest do not either because it is not possible—such as carrion, which is already dead, and blood, which has no life in the first place—or because it is unconditionally prohibited—such as swine. These other types can never be lawful. (Translator)

> Abī Ṭalḥa relates from Ibn ʿAbbās, "It means 'except what you slaughter from any of these [five types of injured animals] while it still possesses its soul.' You may eat these, for they have been duly slaughtered." There are similar reports from Saʿīd ibn Jubayr, Ḥasan al-Baṣrī, and Suddī. (*Tafsīr Ibn Kathīr*)

This verse makes it clear that an animal is only lawful if it is killed through a method that Islamic law has prescribed, and that an animal does not become lawful if strangled or beaten or by its blood merely flowing out, however it may happen. Animals that have "been gored to death" and those that have "been eaten by a predator," for instance, may have their blood flow out from the cutting point, yet the Qur'ān explicitly prohibits these animals. It is evident, then, that an animal does not become lawful merely because its blood flows out, even if it flows from the cutting point. The blood must flow out of the animal by manually and deliberately cutting into the jugular veins, as Allāh has ordered.

Mentioning the Name of Allāh when Slaughtering

The general body (*jumhūr*) of the jurists have concluded that for an act of animal slaughter to be acceptable by Islamic law, it is also obligatory that the slaughterer perform *tasmiya*—that is, mention the name of Allāh—at the time of slaughtering.[22] If one intentionally does not mention Allāh's name, the slaughtered animal is unlawful according to Imām Abū Ḥanīfa, Imām Mālik, Imām Aḥmad, and the overwhelming majority of the jurists. In the event that one forgets it, the slaughter is considered legitimate by the Ḥanafīs and Mālikīs, whether in the case

22 *Tasmiya:* The technical term for "mentioning the name of Allāh," which must be done verbally (not just in the heart) immediately before slaughtering. The words usually used are *Bismillāhi Allāhu Akbar,* meaning "In the name of Allāh; Allāh is the Greatest." However, it is acceptable to say other statements (e.g., *Lā ilāha illa 'Llāh*), as long as Allāh's name, and no one else's, is mentioned. For ease of reading, the English word "mention" and its derivatives will be used for *tasmiya* throughout the text. (Translator)

of voluntary slaughter or hunting. However, the Ḥanbalīs view that forgetting is pardoned only for voluntary slaughter. For hunting, the slaughter is not acceptable if the hunter does not mention the name of Allāh when releasing his arrow (or other projectile) or sending out his hunting dog, regardless of whether he forgets to mention or intentionally abandons doing so. (*Badāʾiʿ al-Ṣanāʾiʿ* 5:46 for Ḥanafī school; *Al-Dhakhīra* 4:134 and *Al-Ṣāwī ʿala ʾl-Dardīr* 2:171 for Mālikī; *Al-Mughnī* 4:11 for Ḥanbalī)

Regarding Imām Shāfiʿī, the notion has become widespread that he views the mentioning of Allāh's name not as obligatory but only *sunna* (*Qalyūbī wa ʿUmayra* 4:245), which would imply that a slaughtered animal is lawful even if the slaughterer purposely does not mention Allāh's name. However, a review of Imām Shāfiʿī's book *Kitāb al-Umm* makes it clear that he did not deem lawful an animal over which the slaughterer intentionally abandons mentioning Allāh's name. He merely deemed lawful that over which the slaughterer *forgets* to mention the name of Allāh. His words are as follows:

> When a Muslim sends out his trained hunting dog or bird, I prefer that he mention the name of Allāh. But if he forgets to mention Allāh's name and his animal then makes a kill, he may still eat the prey. The reason is that if one forgets to mention Allāh's name when manually slaughtering, he may eat the animal, because a Muslim slaughters in the name of Allāh Almighty, even if he forgets to actually utter His name, and because the kill that a Muslim makes with his hunting animal is considered equal to killing the animal by hand-slaughtering it. (*Kitāb al-Umm* 2:227)

Then, later in the book, Imām Shāfiʿī states that if one scornfully abandons mentioning Allāh's name at the time of slaughtering, it is not lawful to eat such a slaughtered animal. Presenting what is unquestionably his view, he writes: "If a Muslim forgets to mention the name of Allāh, his slaughtered animal may be eaten, but if he abandons it scornfully,[23] his slaughtered animal may not be eaten." (*Kitāb al-Umm* 2:131)

---

23 There are two words used that imply the intentional failure to mention Allāh's name (as

Some scholars have stated that the jurists have agreed on this opinion with consensus. The following, for instance, is reported in *Al-Tafsīr al-Maẓharī,* quoting from *Sharḥ al-Muqaddimat al-Mālikiyya:*

> All this is in regard to one who is not disdainful of mentioning Allāh's name. As for one who is disdainful, there is no dispute that it is prohibited to eat his slaughtered animal. Ibn al-Ḥārith and Bashīr have also declared this. In particular, a disdainful person is one who repeatedly abandons mentioning Allāh's name. And Allāh knows best. (*Al-Tafsīr al-Maẓharī* 3:318)

Imām Shāfiʿī's own statements prove that his school does not uphold the unconditional lawfulness of a slaughtered animal over which Allāh's name is intentionally not mentioned. In fact, his view is that a slaughtered animal is prohibited if one abandons mentioning Allāh's name scornfully and disdainfully and makes habit of doing so. That is to say, in effect, that Imām Shāfiʿī restricts the ruling of lawfulness to the case when one *accidentally* fails to mention Allāh's name once or twice; and even in this case, it is still disliked, in his view, to eat such an animal, because he says, "I prefer that he mention the name of Allāh." The Shāfiʿī jurists have therefore said that failing to mention Allāh's name is disliked and, in fact, sinful (*Rawḍat al-Ṭālibīn* 3:205; *Raḥmat al-Umma* 118).

In light of this, we see that the reality is that an animal over which one intentionally does not mention Allāh's name is prohibited in the view of the Ḥanafīs, Mālikīs, and Ḥanbalīs, and also prohibited in Imām Shāfiʿī's view if one does so scornfully and disdainfully and makes doing so a habitual practice. But if it happens by mistake once or twice, his opinion is that it is not prohibited but nevertheless disliked.

This last statement, however, lacks support from the Qur'ān and Sunna, as both the Qur'ānic verses and the Prophetic ḥadīths deem

---

opposed to forgetting to do so): *istikhfāf* and *tahāwun,* which I have translated as "scornfulness" and "disdain," respectively. Though I have consistently maintained separate translations for the two words, they are essentially synonymous in this context. (Translator) mentioning Allāh's name an essential part (*rukn*) of a lawful slaughter. Allāh Almighty says:

> And do not eat of that over which the name of Allāh has not been mentioned [at the time of slaughtering]. For it is indeed sheer deviance. (Qur'ān 6:121)

What statement could be more unequivocal than this noble verse, which is not vague or ambiguous in the least, in showing that an animal over which one has abandoned mentioning Allāh's name is prohibited? It contains an explicit negative command, and a negative command here necessarily means prohibition. But the Qur'ān is not satisfied merely with a negative command. It follows it up with the words "for it is indeed sheer deviance (*fisq*)," eliminating from the matter any remaining smidgeon of doubt. This verse is not the only one that proves that mentioning Allāh's name is an essential part of slaughtering; there are many. Among them are the following:

(1) They ask you about what is permitted to them. Say, "Permitted to you are all wholesome things (*ṭayyibāt*) and [the prey of] the hunting creatures that you have taught, training them as hounds and teaching them from what Allāh has taught you. So eat from what they catch for you. And mention the name of Allāh over it [when sending them out]." (Qur'ān 5:4)

(2) For each faith-community We have appointed a [sacred] rite so that they may mention the name of Allāh over any sacrificial animals that He has provided them. (Qur'ān 22:34)

(3) So mention the name of Allāh over them as they stand in ranks [for sacrifice]. (Qur'ān 22:36)

(4) And cattle over which they do not mention the name of Allāh—which is mere forgery against Him. (Qur'ān 6:138)

(5) And why should you not eat of that over which the name of Allāh has been mentioned? (Qur'ān 6:119)

All of these verses prove, in different ways, that mentioning the name of Allāh at the time of slaughtering is among the most important factors in making eating the meat of an animal lawful. The Qur'ān does not state this in general terms in just one or two verses, but mentions this essential factor independently for slaughtering, hunting, and holiday sacrificing, and it severely rebukes the one who abandons it, deeming his action forgery against Allāh. It also rebukes the one who does not consider as lawful the slaughtered animal over which Allāh's name has been mentioned. All this establishes that mentioning the name of Allāh is one of the most crucial requirements for a lawful slaughter.

Furthermore, there are multiple ḥadīths in which the Messenger of Allāh ﷺ indicates that mentioning Allāh's name is one of the essential elements that must be carried out in order for slaughtered animals or game to be lawful. Consider the following ḥadīths:

(1) It is narrated that Rāfiʿ ibn Khadīj ﷺ said:

> The Messenger of Allāh ﷺ said, "[As long as you use] whatever causes the blood to flow out and the name of Allāh has been mentioned [before slaughtering], you may eat [the animal].'" (*Ṣaḥīḥ al-Bukhārī*)[24]

(2) ʿAbdullāh ibn ʿUmar ﷺ narrates that:

> The Messenger of Allāh ﷺ met Zayd ibn ʿAmr ibn Nufayl[25] at the lower part of [the valley of] Baldaḥ. This was before the Messenger of Allāh ﷺ received revelation. A table spread of food was presented to the Prophet ﷺ, but he refused to eat from it. [He offered it to] Zayd, [who] then said [to the Qurayshite pagans]: "I do not eat from that which you slaughter

24 *Kitāb al-Dhabā'iḥ wa 'l-Ṣayd, Bāb al-Tasmiya ʿala 'l-Dhabīḥa.*

25 Zayd (may Allāh have mercy on him) was a *ḥanīfī,* or follower of the then mostly extinct religion of Abraham ﷺ. He was the father of Saʿīd ibn Zayd, one of the first Muslims, a member of the ten companions promised Paradise by the Messenger ﷺ himself, and the husband of Fāṭima bint al-Khaṭṭāb, the sister of ʿUmar al-Fārūq ﷺ. Zayd died before the first revelation, but the Messenger ﷺ testified that he saw Zayd dragging his clothes about in Paradise. (Translator)

on your [pagan] altars. I do not eat except that upon which the name of Allāh has been mentioned." (*Ṣaḥīḥ al-Bukhārī*)[26]

This is also evidence that the law of Abraham ﷺ prohibited animals over which the name of Allāh is not mentioned.

(3) Jundub ibn Sufyān al-Bajalī ﷺ relates:

> One day [the day of ʿĪd al-Aḍḥā] we offered our sacrifices with the Messenger of Allāh ﷺ, and there were some people who offered theirs before the [ʿĪd] prayer. [After the prayer,] when the Prophet ﷺ turned, he noticed that they had slaughtered before the prayer. So he said, "Whoever slaughtered before the prayer must slaughter another in its place. And whoever did not slaughter before we prayed, let him slaughter in the name of Allāh. (*Ṣaḥīḥ al-Bukhārī*)[27]

(4) ʿAbāya ibn Rifāʿa narrates from his grandfather that the Prophet ﷺ said:

> [As long as you use] whatever causes the blood to flow out and the name of Allāh has been mentioned [before slaughtering], you may eat [the animal]. (*Ṣaḥīḥ al-Bukhārī*)[28]

(5) Abū Thaʿlaba al-Khushanī ﷺ narrates that he asked the Messenger of Allāh ﷺ some questions. The Messenger of Allāh ﷺ answered his question about hunting by saying:

> Whatever you hunt with your bow, mention the name of Allāh [over the arrow before releasing it], and then you may eat [the animal you hit]. And whatever you hunt with your trained dog, mention the name of Allāh [over the dog before sending it out], and then you may eat [the animal it catches]. (*Ṣaḥīḥ al-Bukhārī*)[29]

(6) ʿAdī ibn Ḥātim ﷺ narrates that the Messenger of Allāh ﷺ said:

---

26 *Kitāb Manāqib al-Anṣār, Bāb Ḥadīth Zayd ibn ʿAmr ibn Nufayl.*

27 *Kitāb al-Dhabāʾiḥ wa ʾl-Ṣayd, Bāb Qawl al-Nabiyyi ﷺ "Fa ʾlyadhbaḥ ʿalā ʾsmi ʾLlāh."*

28 *Kitāb al-Dhabāʾiḥ wa ʾl-Ṣayd, Bāb Mā Anhara ʾl-Dam min al-Qaṣab.*

29 *Kitāb al-Dhabāʾiḥ wa ʾl-Ṣayd, Bāb Āniyat al-Majūs.*

"If you send out your trained [hunting] dogs and mention the name of Allāh [over them before sending them], you may eat from what they catch for you." (*Ṣaḥīḥ al-Bukhārī*)[30]

(7) ʿAdī ibn Ḥātim ﷺ relates:

I said, "O Messenger of Allāh, I sent out my [hunting] dog and then found with it another dog [by a dead animal]. I did not know which of them made the catch." He replied, "Do not eat [the animal], as you mentioned the name of Allāh over your dog but not over the other." (*Ṣaḥīḥ al-Bukhārī*)[31]

(8) ʿAdī ibn Ḥātim ﷺ also narrates that the Messenger ﷺ said:

If [your dog] mixes with dogs over which the name of Allāh has not been mentioned and then they catch and kill [an animal], do not eat it." (*Ṣaḥīḥ al-Bukhārī*)[32]

(9) ʿAdī ibn Ḥātim ﷺ also relates:

I said, "O Messenger of Allāh, if one of us gets a hold of some game [while hunting] and does not have a knife with him, may he slaughter using a flint-stone or the [pointed] tip of a staff?" He replied, "Make the blood flow out with whatever you wish and mention the name of Allāh Almighty.'" (*Sunan Abī Dāwūd*)[33]

These passages from the Qur'ān and Sunna show the extent of emphasis and importance placed on mentioning the name of Allāh at the time of slaughtering. Though just one of these references would be sufficient in illustrating that mentioning His name is an essential part of slaughtering, Allāh, the Supreme Lawgiver, clarifies this rule over and over again, on a number of different occasions, and with various styles. This repeated emphasis patently demonstrates that mentioning

---

30 *Kitab al-Dhabā'iḥ wa 'l-Ṣayd, Bāb Mā Jā'a fi 'l-Taṣayyud.*

31 *Kitāb al-Dhabā'iḥ wa 'l-Ṣayd, Bāb Idhā Wajada maʿa 'l-Ṣayd Kalban Ākhar.*

32 *Kitāb al-Dhabā'iḥ wa 'l-Ṣayd, Bāb al-Ṣayd Idhā Ghāba ʿanhu Yawmayni aw Thalāthan.*

33 *Kitāb al-Ḍaḥāyā, Bāb al-Dhabīḥa bi 'l-Marwa.*

Allāh's name at the time of slaughtering is immensely important and is an indisputable requirement for the lawful slaughtering of any animal that we have been permitted to benefit from.

The only time when the obligation of mentioning Allāh's name is relaxed is when the slaughterer forgets. ʿAllāma Jaṣṣāṣ (may Allāh have mercy on him), a great Ḥanafī scholar of the fourth century AH, writes in his *Aḥkām al-Qur'ān:*

> Forgetting to mention Allāh's name does not invalidate one's slaughter because the verse "And do not eat of that over which the name of Allāh has not been mentioned" (6:121) refers to one who does so intentionally, not one who forgets. This is evidenced by the very next words of the verse: "For it is indeed sheer deviance." This deviance is not a characteristic of someone who forgets, as one who forgets is, at the time of forgetting, not held responsible for mentioning the name of Allāh. Awzāʿī narrates from ʿAṭā' ibn Abī Rabāḥ, who narrates from ʿUbayd ibn ʿUmayr, who narrates from ʿAbdullāh ibn ʿAbbās, who relates, "The Messenger of Allāh ﷺ said, 'Allāh has excused my *Umma* (nation) from [errors done by] mistake, and [sins committed because of] forgetting, and [wrongdoing] that they have been coerced to do.'"
>
> Thus, since [one who forgets] is not held responsible for mentioning the name of Allāh, he will have carried out the slaughtering satisfactorily, and failure to mention Allāh's name will not invalidate it. Furthermore, it is not permissible to require him to repeat the slaughtering, since it has irrevocably taken place. This is not like forgetting to recite the opening *takbīr* of a prayer or forgetting to be in a state of purity when performing it, because what comes after these [prerequisite acts] is another mandatory act. But in the case of slaughtering an animal, the possibility of any subsequent mandatory act is precluded because the single act of slaughtering an animal, once done, is irreversibly complete.[34] (Jaṣṣāṣ, *Aḥkām al-Qur'ān* 3:7–8)

34 To understand better the distinction Jaṣṣāṣ makes here between prayer and slaughtering an animal, it is helpful to recognize that slaughtering is the act of taking the life of a living creature, which, once dead, cannot be resurrected to life to be slaughtered again if

Further support for the relaxed requirement of mentioning Allāh's name is a narration that Dāraquṭnī and Bayhaqī have reported on the authority of Ibn ʿAbbās ﷺ, in which the Prophet ﷺ says:

> For a Muslim, that he is a Muslim is sufficient, so if he forgets to mention the name of Allāh when slaughtering, he should mention the name of Allāh afterwards. He should mention the name of Allāh and then eat. (*Naṣb al-Rāya* 2:261)

Ḥāfiẓ Ibn Ḥajar al-ʿAsqalānī mentions this ḥadīth in his *Talkhīṣ al-Ḥabīr,* whereupon he states, "Ibn al-Sakan has determined it to be rigorously authenticated." Some ḥadīth scholars (*muḥaddithīn*) have found it to be defective (*maʿlūl*) because of the narrators Maʿqil ibn ʿAbdillāh and Muḥammad ibn Yazīd ibn Sinān. However, Maʿqil ibn ʿAbdillāh is among Imām Muslim's reliable narrators, and Muḥammad ibn Yazīd ibn Sinān has been declared trustworthy (*thiqa*) by Ibn Ḥibbān, Nufaylī, and Maslama (*Iʿlā' al-Sunan* 17:68). Furthermore, ʿAbd ibn Ḥumayd reports a *mursal* ḥadīth—that is, one that a Follower (*tābiʿī*) narrates from the Messenger ﷺ without mentioning the narrator or narrators in between—from Rāshid ibn Saʿd, in which the Prophet ﷺ says:

> The slaughtered animal of a Muslim is lawful if he mentions the name of Allāh; and [it is lawful] if he does not mention the name of Allāh, as long as he does not abandon intentionally. The same [is true] for hunting.

Imām Suyūṭī also mentions this ḥadīth in *Al-Durr al-Manthūr* (3:42).

These narrations, directly narrated of the Messenger ﷺ [in Ḥadīth terminology called *marfūʿ*], are further supported by one that Bukhārī

---

the act was done incorrectly. For this reason, in the single case that the slaughterer forgets to mention Allāh's name before killing the animal, the animal does not need to be wasted, nor does the slaughterer have to kill another animal in its place. Prayer, on the other hand, is different because it is not a living being that cannot be replaced by another. It is, rather, a collection of actions to which we assign the qualitative name *ṣalāt,* and any such qualitative act, if done improperly, can (and in the case of prayer, must) be redone. (Translator) reports of the great companion Ibn ʿAbbās ﷺ, who said, "Whoever forgets [to mention the name of Allāh], it is not a problem" (*Ṣaḥīḥ al-Bukhārī*).[35] Dāraquṭnī, Saʿīd ibn Manṣūr, and others have said that this ḥadīth has an unbroken chain of transmission (*mawṣūl*), and Ibn Ḥajar al-ʿAsqalānī states, "Its chain of transmission is rigorously authenticated" (*Fatḥ al-Bārī* 9:624).

In comparison to these texts, which prove that mentioning the name of Allāh when slaughtering is obligatory, those that some Shāfiʿīs use to support the contention that mentioning His name is not necessary do not come close in validity or soundness. For example, some of them draw their conclusion from Allāh's words "except that which you duly slaughter," saying that Allāh generalizes slaughtering and does not impose the restriction of mentioning His name; so it would seem that it is not necessary. The response to this is simple: Slaughtering (*dhakāt*) has a specialized meaning in Islamic law, and the texts presented so far prove that it does not take place lawfully except with the mention of Allāh's name. Mentioning His name is therefore included in the legal meaning of slaughtering, just as cutting into the jugular veins is included. Allāh thus refers to slaughtering as a comprehensive concept comprising all of its essential elements (*arkān*) as established by other legal sources. Hence, mentioning Allāh's name is in fact understood in the words "except that which you duly slaughter."

Other Shāfiʿīs have drawn their conclusion from a ḥadīth reported by Imām Bukhārī, in which ʿĀ'isha ﷺ relates the following: "A group of people once said to the Prophet ﷺ, 'Some people have come to us with meat, and we do not know whether the name of Allāh has been mentioned over it or not.' He replied, 'You yourselves mention the name of Allāh over it and eat.'" She then remarks, "They [the ones who offered the meat] had just recently entered into Islam" (*Ṣaḥīḥ al-Bukhārī*).[36]

---

35 *Kitāb al-Dhabā'iḥ wa 'l-Ṣayd, Bāb al-Tasmiya ʿala 'l-Dhabīḥa wa Man Taraka Mutaʿammidan.*

36 *Kitāb al-Dhabā'iḥ wa 'l-Ṣayd, Bāb Dhabīḥat al-Aʿrāb.*

From this ḥadīth, however, one cannot conclude that an animal is lawful when it is known for certain that the slaughterer intentionally did not mention Allāh's name. The intent of this ḥadīth is to teach the correct way to regard the action of a Muslim. In particular, if a Muslim offers meat or some other food, the outward reality is that it has been slaughtered according to the prescribed Islamic way; therefore, the nature of the slaughter should be judged by its outward appearance. We are ordered to think well of every Muslim, so seeking out the slaughtering method is not necessary in this case, as long as it does not become known that the Muslim offering the meat failed to slaughter the animal in accordance with the Islamic method.

In the ḥadīth, the people about whom the questioners ask are indeed Muslims, even though they had recently been non-Muslims, as ʿĀ'isha ﵂ mentions. The Messenger of Allāh ﷺ thus orders that their action be judged by its outward appearance, namely, that they, being Muslims, did in fact mention the name of Allāh when slaughtering. However, it does not follow from this ḥadīth that an animal is lawful if one is certain that its slaughterer intentionally did not mention Allāh's name, since it is self-evident that the question in this ḥadīth only refers to when we do not know for sure whether the Muslim slaughterer mentioned the name of Allāh over the animal or not. In fact, this ḥadīth clarifies the ruling for the situation most Muslims today face concerning the meat found in Muslim markets, for we have not witnessed how exactly the slaughter was performed. However, the situation this ḥadīth addresses—the one in which there is an absence of surety—is worlds apart from when we know for a fact that the slaughterer intentionally and deliberately abandoned the mention of Allāh's name.

Other Shāfiʿīs draw their conclusion from a ḥadīth reported by Abū Dāwūd in his *Marāsīl,* in which the Messenger of Allāh ﷺ says:

> An animal slaughtered by a Muslim is lawful, whether he mentions Allāh's name over it or not, for if he were to mention [someone's name], he would not mention the name of anyone but Allāh. (*Marāsīl Abī Dāwūd* 41)

This ḥadīth is narrated by Ṣalt al-Sadūsī, who is "unknown" (*majhūl*) according to Ibn Ḥazm and Ibn al-Qaṭṭān; he is not known other than by this ḥadīth, and no one has narrated from him other than Thawr ibn Yazīd (*Naṣb al-Rāya*). Its chain of transmission, therefore, is not free of weakness. And even if it were correctly proven to be sound, then it is feasible and fitting to use it as just further proof that the obligation of mentioning Allāh's name is relaxed in the case of forgetting. In this way, the ḥadīth's implication would be in accord with the numerous clear-cut texts that prove that mentioning Allāh's name when slaughtering is mandatory and that an animal over which the slaughterer intentionally does not do so is prohibited.

Because of these strong counterarguments, some Shāfiʿī scholars have given preference in this matter to the view of the general body of the scholars—that mentioning Allāh's name is obligatory. Ḥāfiẓ Ibn Ḥajar al-ʿAsqalānī writes in his *Fatḥ al-Bārī:*

> Imām [Abū Ḥāmid] al-Ghazālī upholds this view in his *Iḥyāʾ ʿUlūm al-Dīn.* He argues that the outward import of the Qur'ānic verse ["And do not eat of that over which the name of Allāh has not been mentioned. For indeed it is sheer deviance" (6:121)] and of the related ḥadīths is that the name of Allāh must, under all circumstances, be mentioned when slaughtering. Furthermore, he argues, the ḥadīths evidencing the permissibility of not mentioning His name can be interpreted either as generally applying in all cases or as a specific reference to one who forgets; it is more fitting, however, to take the latter interpretation so that the collective body of proofs for this issue yield a consistent conclusion that is in line with the outward import of all the relevant sources—namely, that one who forgets to mention Allāh's name is excused, but one who intentionally does so is not. (*Fatḥ al-Bārī* 9:624)

Ḥāfiẓ Ibn Ḥajar does not follow this up with anything further. He mentions these arguments of Ghazālī (may Allāh have mercy on him) in the chapter on the "Slaughtered Animals of Bedouin Arabs." It seems that Ibn Ḥajar was also inclined toward favoring the opinion held by the general body of the scholars, since he mentions Ghazālī's statement at the end of the chapter and declares weak (*ḍaʿīf*) the ḥadīth that he himself presents as possible evidence for permitting an animal over which Allāh's name is not mentioned (*Fatḥ al-Bārī* 9:634).

Religious Status of the Slaughterer

One of the main requirements for a lawful slaughter is that the slaughterer, aside from being fully sane and of the age of discretion, be a Muslim or Kitābī. The jurists have agreed on this requirement, and there is no known dissenting view, to the extent that some scholars have even reported consensus in this matter (*Mawsūʿat al-Ijmāʿ* 2:912, 948). This therefore means that an animal slaughtered by a non-Muslim not from the People of the Book is unlawful and may not be eaten, even if he slaughters the animal as the Muslims do. ʿAllāma Jaṣṣāṣ states, "We know that even if the polytheists mention the name of Allāh when slaughtering, the animals they slaughter may not be eaten" (Jaṣṣāṣ, *Aḥkām al-Qur'ān* 3:6).

There are, however, some contemporaries who have parted from this dominant opinion by strangely limiting this prohibition only to the slaughtered animals of Arab polytheists, and deeming lawful the slaughtered animals of all other non-Muslims, be they idolaters, agnostics, atheists, or fire worshippers. This opinions is erroneous and finds no basis in the Qur'ān or Sunna, nor in the statements of the Righteous Predecessors (*Salaf Ṣāliḥ*). They attempt to support their errant view by doing the following: first, claiming that there is no explicit text from the Qur'ān or Sunna verifying that the slaughtered animals of non-Muslims other than the People of the Book are prohibited; second, applying [incorrectly] the juristic principle that "all things are permissible by default and cannot be declared prohibited except with a proof text" (*Faṣl al-Khiṭāb fī Dhabā'iḥ Ahl al-Kitāb* 19–22).

Though this principle is recognized in Islamic jurisprudence, the one that applies in matters of animal slaughter is that "slaughtered animals are *prohibited* by default and do not become lawful except through some means prescribed by Islamic law." The main proof that this is the applicable principle is the ḥadīth of ʿAdī ibn Ḥātim ﷺ presented earlier:

> I said, "O Messenger of Allāh, I sent out my [hunting] dog and then found with it another dog [by a dead animal]. I did not know which of them made the catch." He replied, "Do not eat [the animal], as you mentioned the name of Allāh over your dog but not over the other." (*Ṣaḥīḥ al-Bukhārī*)

This ḥadīth proves that whenever there is doubt about whether an animal has been properly slaughtered, and both possibilities are equally likely, the animal is prohibited. Thus, it is proof that animals are indeed prohibited by default; for if they were permissible by default, they would not be prohibited in the case of doubt.

Furthermore, the following verse specifically mentions that the animals slaughtered by the People of the Book are permissible:

> And the food of those who have been given the Book is permitted to you. (Qur'ān 5:5)

If the food of all non-Muslims were lawful, He would not have made special mention of the People of the Book. Contrary to what some contemporaries claim, this evidence is not based on the juristically invalid concept of linguistic contraposition (*mafhūm al-laqab*), which would exclude all unmentioned items from a stated ruling and, if applied here, would indeed give us our present conclusion.[37] In fact, the evidence

---

37 A linguistic category of traditional logic's contrapositive reasoning, the concept of *mafhūm al-laqab* is, briefly stated, the notion that a meaning (and, in the context of Islamic jurisprudence, the related ruling) stated in a statement does not apply to anything other than the item(s) or person(s) stated in the text. For example, in the Qur'ānic statement "Muḥammad is the messenger of Allāh" (48:29), by this concept we can conclude that no one but Muḥammad ﷺ is the messenger of Allāh, but this is clearly untrue. Therefore, the Muslim jurists have stated that *mafhūm al-laqab*, as well as all forms of contrapositive reasoning, are invalid in Islamic jurisprudence and are only admissible insofar as they further corroborate

is that any slaughtered animal that is not explicitly mentioned in the verse—namely, that of anyone other than the People of the Book—falls under the default ruling: prohibition.

Therefore, the correct conclusion, upon which the scholars of the *Umma* have agreed over the centuries, is that a slaughtered animal does not become lawful for Muslims unless the slaughterer is a Muslim or a member of the People of the Book. And the "People of the Book" are the Jews and Christians, and none other.

There are some very isolated opinions that regard the Magians (*Majūs*) from among the People of the Book. These are deduced from the statement of the Messenger of Allāh ﷺ: "Do with them [the Magians] as you do with the People of the Book." (*Al-Muḥallā* 7:456)

However, the fact is that this ḥadīth pertains only to collecting the protection tax (*jizya*), which is why it is reported in the Ḥadīth books under this topic. During his caliphate, ʿUmar ibn al-Khaṭṭāb ؓ was hesitant to collect the tax from them. When ʿAbd al-Raḥmān ibn ʿAwf ؓ informed him of this ḥadīth, ʿUmar ؓ went ahead and collected it. Imām Mālik reports, on the authority of Muḥammad ibn ʿAlī, that ʿUmar ؓ sought counsel regarding the Magians' payment of the tax, asking, "What do you think I should do with them?" ʿAbd al-Raḥmān ibn ʿAwf ؓ replied, "I testify that I heard the Messenger of Allāh ﷺ say, 'Do with them as you do with the People of the Book'" (*Muwaṭṭā' Imām Mālik*).[38]

Moreover, the general body of the scholars have concluded from the following verse that the title "People of the Book" (*Ahl al-Kitāb*) is limited only to the Jews and Christians: "[...] or else you would say, 'Indeed the Book was only sent down upon two groups before us'" (Qur'ān 6:156).

Additionally, since the Messenger of Allāh ﷺ explicitly did not regard the Zoroastrians as part of the People of the Book—but instead

a ruling derived from acceptable proof texts. The author mentions this point here to forestall the claim that he arrived at his conclusion using this concept. (Translator)

38 *Kitāb Al-Dhabā'iḥ, Bāb Jizya Ahl al-Kitāb.*

said, "Do with them as you do with the People of the Book"—it is evident that they are not from among them. They are to be treated like the People of the Book only for the collection of taxes.

11

People of the Book: Slaughtered Animals & Related Issues

Regarding the People of the Book, the scholars of the *Umma* have unanimously agreed that their slaughtered animals are lawful. The proof for this is the following verse: "And the food of those who have been given the Book is permitted to you." (Qur'ān 5:5)

The meaning of "food" in this verse is slaughtered animals, according to the unanimous agreement of the scholars. Ḥāfiẓ Ibn Kathīr writes: "[Regarding the word 'food' in the verse] 'And the food of those who have been given the Book is permitted to you,' Ibn ʿAbbās, Abū Umāma, Mujāhid, Saʿīd ibn Jubayr, ʿIkrima, ʿAṭā', Ḥasan, Makḥūl, Ibrāhīm al-Nakhaʿī, Suddī, Muqātil, and Ibn Ḥibbān have said, 'It means slaughtered animals.' This ruling is unanimously agreed upon by the scholars: their slaughtered animals are lawful for Muslims because they believe that slaughtering for other than Allāh is prohibited, and because they do not mention any name except the name of Allāh over the animals they slaughter—even though they do espouse certain beliefs above which Allāh is greatly exalted." (*Tafsīr Ibn Kathīr* 2:19)

But are the requirements for slaughtering done by Muslims—cutting the jugular veins, using a sharp tool, and mentioning the name of Allāh—also required for slaughtering done by the People of the Book? Arriving at the correct answer requires an analysis with a look at what some contemporary scholars have claimed, namely, that the slaughtered animals of the People of the Book are lawful regardless of what method they use. We will treat this topic in two parts: (1) whether it is obligatory that the People of the Book slaughter animals with the Islamically prescribed method of cutting the jugular veins with a sharp tool, and (2) whether it is necessary that they mention the name of Allāh when slaughtering.

Slaughtering the Animals with the Islamic Method

The general body of the jurists have agreed that the slaughter of a Kitābī is only lawful if he slaughters by cutting into the jugular veins with a sharp tool. This is a fact established by conclusive proofs that we will mention shortly, Allāh willing.

But first let us take a look at what some contemporary scholars claim. They assert that the animal slaughtered by a Kitābī is lawful, regardless of the method he uses to kill it. They base this on the general import of the verse, "And the food of those who have been given the Book is permitted to you" (Qur'ān 5:5). They also cling to the statement of Qāḍī Ibn al-ʿArabī (may Allāh have mercy on him), wherein he says:

> I was once asked the following question about a Christian who twists off the neck of a chicken and then cooks it: "May a Muslim eat this meat with him or may it be taken from him as lawful food?" This is issue number eight. I responded: "It may be eaten because it is his food and the food of his priests and monks. Even if this method is not a legitimate way of slaughtering in our view, Allāh Most High has categorically made their food lawful. Everything they view as lawful in their religion is lawful in our religion, except what Allāh specifically deems prohibited." (Ibn al-ʿArabī, *Aḥkām al-Qur'ān* 2:554)

This unusual statement of Ibn al-ʿArabī, however, completely contradicts the principle that he himself mentions in the same book about one page earlier. His words there are as follows:

> If it is asked, what about what [the People of the Book] eat that has been slaughtered improperly, such as by strangulation or smashing of the head? then the answer is that it is considered to be carrion (*mayta*) and is therefore prohibited (*ḥarām*) according to an explicit proof [from the Qur'ān]. Even if they eat it, we may not. This is as with swine: it is lawful for them and one of their foods, but it is prohibited for us. The same ruling applies here. And Allāh knows best. (Ibn al-ʿArabī, *Aḥkām al-Qur'ān* 2:553)

There is an obvious contradiction in Ibn al-ʿArabī's two statements. And whenever there is a contradiction between two statements, it is only appropriate to accept the one that is corroborated by other sources and supported by the uninterrupted practice of the *Umma,* not the anomalous one opposed by many strong proofs. Presented in the remaining pages of this section are seven points that refute Ibn al-ʿArabī's first statement in particular and, more generally, the claim that the slaughtered animal of a Kitābī is lawful irrespective of the slaughtering method used.

1. Allāh says:

> Forbidden to you is [the eating of] carrion and blood and the flesh of swine and that over which other than [the name of] Allāh has been invoked and what has been strangled and what has been beaten to death and what has fallen to [its] death and what has been gored to death and what has been eaten by a predator—except that which you duly slaughter [before it dies]. (Qur'ān 5:3)

Here Allāh categorically prohibits "what has been strangled" and "what has been beaten to death." This includes everything that dies by being asphyxiated—strangulation, suffocation, and so on—or by being fatally beaten. Thus, if anyone tries to infer from the generality of verse

5:5 that an animal strangled or beaten to death by a Kitābī is lawful, then he must also concede that swine slaughtered by a Kitābī is lawful, as it is also part of "the food of those who have been given the Book." But if verse 5:3 is used to prove that swine is prohibited, then the very same verse likewise prohibits an animal strangled or beaten to death. There is no way to separate the two.

With this in mind, it's obvious that if verse 5:3 specifically excludes swine as part of the general category of the "food of those who have been given the Book," then it equally excludes the animal strangled or beaten to death. In fact, animals strangled and beaten to death are excluded even more so. This is because swine is lawful in Christianity, but an animal strangled or beaten to death is actually prohibited according to the principles of their religion, as we will see shortly, Allāh willing. Thus, if the food that is lawful for them—swine—is an exception to "the food of those who have been given the Book," then the food that is prohibited by their own religious principles—what is strangled or beaten to death—is an exception all the more.

2. A well-established principle of jurisprudence (*fiqh*) and linguistics (*lugha*) is that whenever a ruling is verbally given for a derived noun, the root of derivation of that noun is a cause for that ruling. For example, if we say, "Honor the knowledgeable ones," the order is to give honor to the "knowledgeable ones" (*ʿulamāʾ*), which is the derived noun. The root of derivation, "knowledge" (*ʿilm*), is therefore the cause for being honored. This is obvious. By the same token, since the ruling of prohibition in the verse 5:3 is given for "what has been strangled" and "what has been beaten to death," then strangulation and beating are causes for this ruling. This shows that whenever an animal is strangled or beaten to death, the ruling of prohibition holds and the religion of the strangler or beater is entirely irrelevant. An animal is prohibited if strangled or beaten to death, whether the doer of the action is a Muslim or Kitābī.

3. The established import of the verse "And the food of those who

have been given the Book is permitted to you" is that the People of the Book are like Muslims in their laws of animal slaughter. If, however, we were to accept Ibn al-ʿArabī's first statement—that whatever a Kitābī slaughters is lawful, regardless of how he kills the animal—then we would implicitly be admitting that the People of the Book are superior to Muslims. In effect, we would be saying that if a Kitābī slaughters an animal using an Islamically unlawful method (such as strangling and beating), his animal is lawful and pure, but if a Muslim slaughters using the same method, his animal is prohibited. This is inconceivable and intuitively absurd.

4. Since it is undisputed among the scholars of the *Umma* that all non-Muslims are in reality of one group, it seems to follow that the slaughtered animals of the People of the Book should be prohibited just like that of the rest of the non-Muslims. But Islamic law distinguishes the People of the Book from the rest of the non-Muslims in matters of animal slaughter and marriage, as their laws pertaining to both are similar to those of Islam. They used to conform to many of the same requirements that Islam obligated on the Muslims. These rules are present, even till this day, in their sacred scriptures, despite the many corruptions they presently contain.

Consider the following selections from their scriptures. In the Old Testament, it is reported in the Book of Leviticus:

> The fat of an animal found dead or torn by wild animals may be used for any other purpose, but you must not eat it. (Leviticus 7:24)

The Book of Deuteronomy states:

> The blood of your sacrifices must be poured beside the altar of the Lord your God, but you may eat the meat. Be careful to obey all these regulations I am giving you, so that it may always go well with you and your children after you, because you will be doing what is good and right in the eyes of the Lord your God. (Deuteronomy 12:27–28)

All Jews and Christians acknowledge these two books.

As for the scriptures that only Christians accept, the following is reported in the Book of Acts:

> It seemed good to the Holy Spirit and to us not to burden you with anything beyond the following requirements: You are to abstain from food sacrificed to idols, from blood, from the meat of strangled animals and from sexual immorality. (Acts 15:28–29)

In another place in the same book, it is stated:

> As for the Gentile believers, we have written to them our decision that they should abstain from food sacrificed to idols, from blood, from the meat of strangled animals and from sexual immorality." (Acts 21:25)

Also, St. Paul, who is one of the main leaders of the Christians and an apostle according to their claim, writes in his First Epistle to the Corinthians:

> No, but the sacrifices of pagans are offered to demons, not to God, and I do not want you to be participants with demons. You cannot drink the cup of the Lord and the cup of demons too; you cannot have a part in both the Lord's table and the table of the demons. (1 Corinthians 10:20–21)

It is worth mentioning that St. Paul was the one who, in spite of the teachings of Jesus ﷵ, abrogated nearly all the commands in the Torah that pertained to the Christians. Nevertheless, he still left intact the commands pertaining to slaughtering animals. He prohibited strangled animals and mandated that slaughtering be for God. Hence, it is clear that the original Christian laws for animal slaughter remained as they had been for the Jews.

The exclusively Jewish scriptures are also replete with detailed laws pertaining to animal slaughter. The following is reported in the Mishnah, the primary source of sacred law for Jews:

> If he slaughtered with a hand-sickle or with a flint or with a reed, what he slaughters is valid. All may slaughter and at any time and with any

> implement excepting a reaping-sickle or a saw or teeth or the finger-nails, since these choke. (Mishnah 513–514)

The translator, Dr. Herbert Danby, comments in a footnote under this passage that according to the Jewish tradition given to Moses عليه السلام on Mount Sinai, "five things must be avoided in cutting the throat in order that the slaughtering shall be valid: (a) there must be no delay, but the knife must be kept continually moving backwards and forwards; (b) no pressure may be exerted; (c) there may be no thrusting or digging in of the knife under the skin or between the gullet and windpipe; (d) the knife may not be allowed to slip beyond a certain area of the throat—from the large ring of the windpipe to the upper lobe of the inflated lungs; (e) the gullet or windpipe must not be torn out of position in the course of slaughtering" (Mishnah 513).

The six passages presented above are from books that the Jews and Christians view as holy texts and the primary sources of their religions' sacred laws. What we have quoted verifies the following four points:

(a) Animals strangled and beaten to death are prohibited in Judeo-Christian law, just as they are in Islamic law.

(b) The laws of the People of the Book require that the slaughter be for Allāh—in other words, with the mention of Allāh's name—as is evident from St. Paul's letter to the Corinthians.

(c) When Ibn al-ʿArabī gave his legal opinion that the chicken whose neck a Christian twisted off is lawful (assuming that he actually wrote this passage, given that it blatantly contradicts another passage in the very same book), he gave it only under the impression that a strangled animal is lawful for the Christians, since he says, "Everything that they view as lawful in their religion is lawful in our religion." However, it is now clear that the text of the Christian scriptures explicitly prohibits strangled animals, as we have verified from the Book of Acts 15:28 and 21:25. Had Ibn al-ʿArabī known this fact, he would not have given this opinion.

(d) In light of all this, Ibn Kathīr's statement, previously quoted, is correct: "This ruling is unanimously agreed upon by the scholars: their slaughtered animals are lawful for Muslims because they believe that slaughtering for other than Allāh is prohibited, and because they do not mention any name except the name of Allāh over the animals they slaughter—even though they do espouse certain beliefs above which Allāh is greatly exalted" (*Tafsīr Ibn Kathīr* 2:19).

5. If we were to say that an animal strangled or beaten to death at the hands of a Christian is lawful, this naturally leads to a very absurd conclusion, in view of the texts quoted above. If a Muslim strangles or beats the animal to death, the animal is prohibited in Islam; if a Christian does it, the animal is prohibited in Christianity, but *we* then say that this Christian's animal is lawful for Muslims, even though it is prohibited for Christians. By this logic, an animal strangled at the hands of a Muslim is prohibited for Muslims, but an animal strangled by a non-Muslim is somehow lawful for Muslims. So it is as though the strangler's (or beater's) status as a non-Muslim is actually a special quality that justifies his actions. This nonsensical conclusion naturally results if we say that that which a Kitābī kills is lawful for Muslims, even if he does so in an Islamically unlawful way. Anything that leads to such a baseless conclusion is itself baseless.

6. There are two things that distinguish Jews and Christians from the rest of the non-Muslims in their relationship with Muslims: lawfulness of eating their slaughtered animals and lawfulness of marrying their women.

There is no dispute that marrying a woman from the People of the Book is only lawful if all the mandatory conditions of marriage dictated by Islamic law are fully observed. If a Muslim man marries a woman from the People of the Book in a manner contrary to that of the Muslims—such as marrying one of unmarriageable kin (*maḥram*), or marrying without any witnesses, or marrying without the audible offer and acceptance that Islam requires—no one could maintain that such a marriage is lawful. Therefore, the lawfulness of marrying women from the People of the Book requires that the marriage contract and ceremony be carried out in the manner prescribed by Islamic law. If they are not, then it is incorrect to infer from the verse "And [also permitted to you in marriage are] chaste women from among those who have been given the Book before you" (Qur'ān 5:5) that such a marriage is lawful.

Similarly, how can we say that their slaughtered animals are lawful if they do not slaughter them in an Islamically acceptable manner? If they slaughter an animal in a way not prescribed by Islamic law, such as strangulation or beating, then how can one infer from the verse "And the food of those who have been given the Book is permitted to you" that it is lawful, seeing as both this ruling and the ruling for marrying their women are presented in the Qur'ān together in the exact same context?

7. Because the prohibition of carrion, of what is strangled, and of what is beaten to death, is established by decisive (*qaṭʿī*) proof texts, the jurists have agreed with consensus that they are forbidden, even if the strangler or beater is from the People of the Book. And, from what I know, no one has supported the lawfulness of an animal strangled or beaten by a Kitābī, except Ibn al-ʿArabī in his latter statement. But as we have already seen, this statement totally contradicts another one only about one page before it in the same book. Can the concrete texts of the Qur'ān and Sunna be discarded merely because of the one isolated opinion put forward by Ibn al-ʿArabī—even though it conflicts with another statement of his own and is based on his understanding that a strangled animal is lawful in Christianity, while in fact the Christian scriptures themselves clearly show that this is not the case?

Even if we overlook that his first statement is contradicted and grant that it is his own individual opinion, this is still a very unusual opinion. The sound proofs from the Qur'ān and Sunna embraced by the general body of scholars refute it. It is not right to give this opinion any consideration in a serious issue such as this. In dealing with animal slaughter, whenever there is doubt between a ruling of lawfulness and one of prohibition, the doubt is resolved in favor of prohibition. If this is so in the case of doubt, what about when the corpus of decisive proof texts and the agreement of the scholars favors prohibition?

Thus, the correct conclusion is that animals slaughtered by the People of the Book are not lawful unless they slaughter them with the method prescribed by Islamic law: making the blood flow out by cutting into the jugular veins. It is not lawful if they kill them by strangulation, beating, or any other method prohibited by Islamic law.

Mentioning Allāh's Name when Slaughtering

The second major issue regarding the People of the Book is whether, for their slaughtered animals to be lawful, it is necessary that they mention the name of Allāh when slaughtering. The scholars have different views about this; there are three major ones.

The first is that mentioning Allāh's name is required equally of a Muslim and a Kitābī. This is according to the Ḥanafī and Ḥanbalī schools. Ibn Qudāma al-Ḥanbalī (may Allāh have mercy on him) writes:

> Intentionally mentioning the name of Allāh is required of every slaughterer, whether he is a Muslim or Kitābī. If a Kitābī intentionally does not mention Allāh's name, or if he mentions a name of someone or something other than Allāh, his slaughtered animal is not lawful. This is the view reported of ʿAlī, and Nakhaʿī, Shāfiʿī,[39] Ḥammād, Isḥāq, and the People of Opinion (*Aṣḥāb al-Raʾy*) hold this view. (*Al-Mughnī* 11:56)

The Ḥanafī scholar Kāsānī states in his *Badāʾiʿ al-Ṣanāʾiʿ*:

39 This is what Ibn Qudāma reports of the Shāfiʿī school. It has become commonly thought that in this school, mentioning Allāh's name is not necessary, even for Muslims—and so how can it be necessary for the People of the Book? However, Imām Shāfiʿī did not teach that a slaughtered animal is lawful if the mention of Allāh's name is abandoned scornfully and disdainfully. And outwardly, a disbeliever who abandons the mention of Allāh's name does so out of disdain. Hence, in this sense, Imam Shāfiʿī's view is that if a disbeliever does not mention Allāh's name, his slaughtered animal is not lawful. And Allāh knows best.

The slaughtered animal of a Kitābī may be eaten if the slaughtering is not personally witnessed and the slaughterer is not heard reciting anything, just as it may be eaten if he is seen and heard mentioning only the name of Allāh. This is because if no one hears him say anything, he is assumed to have mentioned the name of Allāh and that of no one else—in order to accord him the benefit of the doubt, as would be done for a Muslim slaughterer.[40] If he is heard pronouncing the name of Allāh, but by "Allāh" he means the Messiah [Jesus] ﷺ, the scholars have said that his animal may be eaten because he outwardly mentions what the Muslims mention. However, if he specifically utters something like "In the name of Allāh, who is one of three," then it is not lawful. It is reported that ʿAlī ؓ was once asked about the slaughtered animals of the People of the Book, who at that time used to mention what the Muslims mentioned. He replied, "Allāh has made their slaughtered animals lawful, and He knows best what they mean by what they say."

Furthermore, if he is heard mentioning the name of the Messiah ﷺ alone or the name of Allāh and the name of the Messiah together, his slaughtered animal may not be eaten. This is what is reported of ʿAlī ؓ, and no one else has reported of him anything to the contrary. (*Badāʾiʿ al-Ṣanāʾiʿ* 5:47)

The second view is that it is not obligatory for a Kitābī to mention the name of Allāh when slaughtering an animal, and so the animal would be lawful if he remains silent and does not mention anything; but if he mentions a name other than that of Allāh, such as that of Jesus ﷺ, his slaughtered animal is not lawful. This is the view of the Mālikī school. Dardīr writes in *Al-Sharḥ al-Ṣaghīr:*

It is mandatory when slaughtering [to mention the name of Allāh], in whatever form, such as "In the name of Allāh" (*Bismi 'Llāh*), "There is no one worthy of worship except Allāh" (*Lā ilāha illā 'Llāh*), "Glorified

40 This rule holds only if it is generally known that the People of the Book of the time mention the name of Allāh when slaughtering. However, if they are commonly known not to mention His name, then this does not apply.

> be Allāh, free of all defects" (*Subḥāna 'Llāh*), or "Allāh is the Greatest" (*Allāhu Akbar*). But this is not for a Kitābī. It is not necessary for him to mention Allāh's name when he slaughters; nevertheless, it is required that he not mention the name of anyone else [besides Allāh] who he believes to be divine. (*Al-Sharḥ al-Ṣaghīr* 2:170–171)

Finally, the third view is that mentioning Allāh's name is not obligatory for a Kitābī and that his slaughter is lawful even if he mentions a name other than Allāh's. According to what Ibn Qudāma reports in his *Al-Mughnī,* this view is reported of ʿAṭā', Mujāhid, and Makḥūl.

When we think about the relevant proof texts, we must conclude that the first opinion is the dominant one. This is because Allāh says: "And do not eat of that over which the name of Allāh has not been mentioned" (Qur'ān 6:121).

The usage of the passive voice here—"the name of Allāh *has not been mentioned"*—is unconditional and is clear proof that abandoning the mention of Allāh's name when slaughtering an animal is prohibited, whether the slaughterer is a Muslim or Kitābī. We see the same thing in the verse in which Allāh enumerates the prohibited foods: "And [forbidden to you is] that over which other than [the name of] Allāh *has been invoked"* (Qur'ān 5:3). Here again the usage of the passive voice includes when the slaughterer is a Muslim or Kitābī. We see the same yet again in another part of the same verse: "And [forbidden to you is] what *has been slaughtered on any altar"* (Qur'ān 5:3).

We observed earlier that both the Jews and the Christians used to slaughter their animals in the name of Allāh. We have also seen that St. Paul, in his first letter to the Corinthians, prohibited to the Christians animals that other peoples had slaughtered because those animals had been "offered to demons, not to God" (1 Corinthians 10:20–21). It is because of such insistence on slaughtering in the name of Allāh that the animals slaughtered by the People of the Book were made lawful, as noted in Ḥāfiẓ Ibn Kathīr's statement (p. 51). Therefore, if they abandon mentioning Allāh's name, or if they mention other than Allāh's name, the key prerequisite by which their slaughtered animals have been made lawful is missing, and so the ruling of prohibition returns. Also, most of the sources that we have seen prohibiting animals strangled or beaten to death by a Kitābī also treat the topic of mentioning Allāh's name. However, the issue of abandoning the mention of Allāh's name is not quite as straightforward as the issue of strangled and beaten animals. As the three varying opinions that we have just seen demonstrate, the prohibition of an animal over which a Kitābī does not mention Allāh's name is subject to scholarly judgment (*ijtihād*). However, there is no dispute among the recognized Imāms about animals strangled or beaten to death (and the statement of Ibn al-ᶜArabī does not deserve any consideration that would make the matter a point of dispute).

At any rate, the correct and dominant opinion, supported by clear proof texts, is that animals slaughtered by the People of the Book are lawful only if they observe all the rules of slaughter laid down in the Qur'ān and Sunna. When permission to eat their slaughtered animals was revealed, they were known to do this. And Allāh knows best.

Other Related Issues

Animals Slaughtered by Materialists, Atheists, and Nominal Kitabıs

For animals slaughtered by the People of the Book to be lawful, it is also required that the slaughterer truly be a Christian or Jew who believes in the fundamental principles of his religion, even if those principles—such as the doctrines of the Trinity and Atonement and the belief that the corrupted Torah (*Tawrāt*) and Gospel (*Injīl*) are in their original form—openly clash with Islam. Allāh named them the People of the Book despite that they persisted in believing in falsehood when the Qur'ān was revealed. In several places, the Qur'ān clearly notes this persistence of theirs: (1) And the Christians say, "The Messiah is the son of Allāh" (Qur'ān 9:30); (2) "Truly they have disbelieved who say that Allāh is one of three" (Qur'ān 5:73); (3) "The Jews say, 'ᶜUzayr is the son of Allāh'" (Qur'ān 9:30); (4) "They altered the [revealed] words from their context" (Qur'ān 5:13).

ʿAllāma Jaṣṣāṣ states:

> ʿUbāda ibn Nasī reports, on the authority of Ghadīf ibn al-Ḥārith, that ʿUmar ibn al-Khaṭṭāb's deputy wrote to him and asked him concerning some people from Sāmira who read the Torah and kept the Sabbath but did not believe in resurrection, "What do you think of them?" ʿUmar wrote back to him saying that they were indeed a denomination of the People of the Book. (Jaṣṣāṣ, *Aḥkām al-Qur'ān* 2:323)

Based on this, it is not required that a Kitābī believe in the pure oneness of Allāh (*tawḥīd*) as Muslims do, nor that he believe that the present Torah and Gospel are corrupted or that the laws of Moses ﷺ and Jesus ﷺ have been abrogated in favor of the law of Muḥammad ﷺ. It is sufficient that he believe in the basic doctrines that the Jews and Christians believe in—those that distinguish them from people of other creeds.

At the same time, however, to regard a person from the People of the Book, it is not enough that he have a Christian or Jewish name or that he be counted as a Christian or Jew in the official statistical records. His beliefs must truly be Christian or Jewish beliefs. In this day and age, especially in the Western nations, there is an innumerable number of people whose names are like those of Christians and Jews, and who may be recorded statistically as Christians and Jews, but are in fact atheists or materialists who do not even believe in the existence of a creator of the universe, let alone the other Judeo-Christian doctrines. In fact, many of them mock religion. Such people are not Christians or Jews, so it is not permissible to consider them from the People of the Book. Hence, their slaughtered animals are not lawful.

The proof for this is clear. The People of the Book are distinguished from the rest of the disbelievers because of their belief in the existence of Allāh and in the messengers and the divine books revealed to them. Therefore, one who outright denies his belief in the existence of Allāh or in the coming of messengers or in the divine revelation of books cannot conceivably be considered among the People of the Book. It is reported that ʿAlī ibn Abī Ṭālib ﷺ gave this very ruling for the Christians of Banū Taghlab. Jaṣṣāṣ writes:

> Muḥammad ibn Sīrīn reports from ʿUbayda, who said, "I asked ʿAlī about the slaughtered animals of the Christians from a certain group of Arabs. He replied, 'It is not lawful because they have not adhered to anything of their religion except the consumption of alcohol!'" (Jaṣṣāṣ, *Aḥkām al-Qur'ān* 2:323)

This means that these people did not believe in the Torah, the Gospel, or any of the other basic doctrines of their religion. Thus, it was not possible to count them among the People of the Book merely because of their nominal association with Christianity.

Nevertheless, this rule only applies to a person if it is *confirmed* that he does not believe in Allāh or the messengers or the divinely revealed books. If he is a Christian by his name and his outward appearance, then it is permissible to regard him as a Christian, as long as nothing on his part betrays that his true beliefs are otherwise.

Ruling When the Slaughterer is Unknown

When the slaughterer and the method he uses are unknown, then one of the following four scenarios must apply:

1. When the country is Muslim—that is, the majority of the residents are Muslims—then the meat found in the stores is lawful to eat, even if we do not know the slaughterer himself or whether or not he mentioned the name of Allāh when slaughtering. This is because everything found in Muslim countries is assumed to have been produced in accordance with Islamic law, since we are commanded to think well of Muslims. The basis of this is the ḥadīth reported by ʿĀ'isha ﷺ, wherein she says:

> A group of people once said to the Prophet ﷺ, "Some people have come

> to us with meat, and we do not know whether the name of Allāh has been mentioned over it or not." He replied, "You yourselves mention the name of Allāh over it and eat."

ʿĀ'isha ﷺ then says, "They [the ones who offered the meat] had just recently entered into Islam." (*Ṣaḥīḥ al-Bukhārī*).[41]

Ḥāfiẓ Ibn Ḥajar al-ʿAsqalānī writes in his explanation of this ḥadīth:

> Ibn al-Tīn states: "There is no legal burden [upon Muslims] to verify whether Allāh's name has been mentioned if a slaughtering is done by someone other than them and they are ignorant of whether Allāh's name was mentioned or not. The slaughtering is judged unacceptable only if it surfaces with certainty that the slaughterer did not mention Allāh's name. It is also possible that this ḥadīth means that if the meat comes from someone whose slaughtered animal becomes lawful if he does in fact mention the name of Allāh, then mentioning His name before eating makes the meat lawful if you do not know whether His name was mentioned at the time of slaughtering. We can gather from this understanding that all meat found in the Muslim markets is judged to be lawful, as is meat slaughtered by Muslim Bedouin Arabs, because for the most part, they know about the requirement of mentioning Allāh's name. Ibn ʿAbd al-Barr has firmly settled on this conclusion. (*Fatḥ al-Bārī* 9:635–636)

Moreover, ʿĀ'isha's words "They had just recently entered into Islam" indicate that it was feared that those who offered the food were unaware of the requirement of mentioning Allāh's name at the time of slaughtering. Despite this, the Messenger of Allāh ﷺ permitted eating their meat because a Muslim's action, even if he is somewhat ignorant, is assumed to be done correctly, so long as no one ascertains with certainty that he performed the action incorrectly. This is the meaning that Imām Bukhārī indicates by heading this ḥadīth with

41 *Kitāb al-Dhabā'iḥ wa 'l-Ṣayd, Bāb Dhabīḥat al-Aʿrāb.*

"The Slaughtered Animals of Bedouins Arabs and Others Like Them." Ibn Ḥajar also notes that Imām Nasā'ī's narration of this ḥadīth states that the people who presented the meat were indeed Bedouins. And Bedouins were generally quite ignorant.

2. The second scenario is when the majority of a country's residents are non-Muslims other than the People of the Book. In this case, the meat for sale in the stores is not lawful for Muslims. Any given meat is only lawful when it is confirmed with certainty, or with the dominant belief of the people, that that particular meat has been slaughtered by a Muslim or a Kitābī with the method prescribed by Islamic law.

3, The same ruling applies when the population of a country is made up of a mix of Muslims, polytheists, Zoroastrians, and others, because doubtful meat is not lawful until one verifies that it really is lawful. The proof for that is in the ḥadīth of ʿAdī ibn Ḥātim ﷺ, mentioned several times earlier, in which the Messenger of Allāh ﷺ prohibited game that stray dogs upon which Allāh's name had not been mentioned participated in hunting down.

4. Finally, when the People of the Book make up the majority of a country's populace, the ruling is like the one for Muslim countries. However, if it is known with certainty, or it is the dominant belief of the people, that the People of the Book in a given country generally do not slaughter animals according to the prescribed method, then eating any given meat is not lawful until one determines that it has been slaughtered properly. And this is the case with most western countries today, as we will discuss further in the coming pages, Allāh willing.

III

Modern-Day, Automated Methods of Animal Slaughter

TODAY'S INCREASINGLY burgeoning population and correspondingly tremendous dietary demands have given rise to the use of automated machines for slaughtering animals. Immense slaughterhouses and butcheries have been established, with daily outputs reaching well into the thousands of animals. It is therefore necessary to gain an understanding of the related Islamic rulings. These machines employ different methods for different animals. The method for slaughtering chickens, for example, is different from that for slaughtering cows and sheep. Let us take a closer look.

Method for Slaughtering Chickens

Overview

The very commonly used method for slaughtering chickens—I have observed it myself in Canada, South Africa, and Réunion Island (located off the eastern coast of Madagascar)—is that one very large machine undertakes all the stages of slaughtering and production. The chickens enter the machine from one side, and refined meat comes out packaged from the other end. This electrically powered machine also carries out all of the stages in between: slaughtering the chickens, plucking their feathers, removing their innards, then cleaning the meat, cutting it into pieces, and finally packaging it.

This machine consists of a long metal rod extending across a large hall from one wall to the other. Along this rod are many hangers, whose hooks face down toward the ground. Hundreds of live chickens are delivered in large trucks. Each chicken is then hung by its legs, such that its legs are attached to the hook of a hanger and the rest of its body hangs upside down, its throat and beak facing the ground. The hangers, with the chickens attached, move down the rod. Then they reach an area into which cold water pours from above in the form of a small waterfall. The hangers pass the chickens through this cold water in order to clean them of dirt and other impurities. In some instances, the water contains a current of electricity that stuns the chickens. Then the hangers come to another area, at the bottom of which is a knife rotating at an extremely high speed. This rotating knife is placed such that the necks of the chickens can easily reach it. So when the hangers reach this area, they travel around the rotating blade in a crescent-shaped pattern, allowing a great number of chickens' necks to make contact with the blade's edge all at once. The knife automatically cuts every throat that hits it cleanly. After the machine has finished cutting the throats of that batch of chickens, the hangers proceed forward.

Soon after, they pass through an area into which, once again, water is poured from above. But this time the water is hot, with the purpose of removing the chickens' feathers. There are then other stages—their intestines are removed, they are purified and cut into pieces, and so on—which are all carried out by the same machine. But we will leave these stages aside as they are outside of the actual slaughtering process, which is the purpose of this book. One side observation that is worth mentioning, however, is that this electric machine continues to run throughout the day, not stopping except in special situations.

There are four parts of this process that require examination through the lens of Islamic law: (1) the passage through the cold water containing the electrical current; (2) the cutting of the chickens' throats with the rotating knife; (3) how the obligation of mentioning Allāh's name can be fulfilled when this automated system is used; and (4) the passage through the hot water.

The Cold Water and the Rotating Knife

Passing the chickens through the cold water before cutting their throats is not used in all slaughterhouses. In fact, many do without it. If the water does not contain any electrical current, it does not have any bearing on the slaughtering of the animal. But if it does, then it usually does not cause the animal to die, but it does send a shock to its brain. We should note that the blood of a slaughtered animal that has first been stunned—even if stunning the animal does cause constriction of the heart—generally does not exit the body in as great a quantity as it does from an animal slaughtered without having been stunned. Nevertheless, stunning itself is usually not what kills the animal. However, if it happens that running the animal through the cold water does kill it—either because the water is too cold or the electrical current is too strong—then it is not permissible to eat it, even if its neck is later cut in the prescribed manner. Therefore, it is essential to assure that the cold temperature of the water and the current of electricity are not so intense that they cause the animal's death. This stage must be carefully overseen so that no dead animals emerge from the water. In any case, it would be best to simply remove this phase entirely, so as to avoid any doubtful situations.

As for the rotating knife, it somewhat resembles a hand mill and has a razor-sharp edge; it spins at a rapid and constant rate. The chickens' necks pass along its edge with their throats facing it, so that their throats are cut automatically. The knife seems to cut the jugular veins of the chickens. But it is quite possible that a chicken, for one reason or another, moves around on the hanger such that its neck does not meet the edge of the rotating knife. If this occurs, one of two things is very possible: either its neck is not cut at all, or only a small part of it is cut, creating doubt as to whether the jugular veins are in fact cut or not. In either case, a lawful slaughtering does not take place.

Mentioning the Name of Allāh

Mentioning Allāh's name is very difficult to do when using this automated system. The first problem is determining who the slaughterer is, because mentioning Allāh's name is required of the slaughterer himself. It is not permissible for one person to mention the name of Allāh and another to slaughter.

So the question is, who really is the slaughterer in this system? We might surmise that the one who turns on the machine in the beginning is regarded as the slaughterer, because, generally speaking, actions and processes carried out by electric machinery are attributed to those who run them. A machine does not have any intellect such that an action might be attributed to it; the action must be attributed to its user who does the action by means of the machine. But the problem here is that whoever turns on this machine at, for example, the start of the day only starts it up one time. It then continues to run throughout the working hours, and sometimes around the clock, cutting the necks of thousands of chickens. So if the one who turns it on mentions the name of Allāh only once—when starting it up at the beginning of the day—does this one mention suffice for the thousands of chickens slaughtered throughout the day? The Qur'ānic statement answers this: "And do not eat of that over which the name of Allāh has not been mentioned" (Qur'ān 6:121). From this verse, it is clear that every animal requires an independent mention, after which it is to be slaughtered immediately. In accordance with this principle, the jurists have extrapolated the following rulings:

1. For voluntary slaughter, one of the requirements is to specify the

animal to be slaughtered by mentioning the name of Allāh. Based on this, we derive the ruling for when the slaughterer mentions Allāh's name and slaughters an animal, and then slaughters another thinking that the first mention suffices for both animals: the second animal may not be eaten, for it is necessary to make a fresh mention for every animal separately. (*Al-Fatāwā al-Hindiyya* 5:286)

2. If one lays down a sheep to slaughter it, takes a knife, and mentions Allāh's name, but then leaves it and slaughters another sheep, having intentionally not mentioned Allāh's name over it, it is not lawful. Thus is related in *Al-Khulāṣa.* (Ibid. 5:288)

3. If one lays down a sheep to slaughter it and mentions the name of Allāh over it, but then speaks to someone, drinks some water, sharpens his knife, eats a bit of food, or does some similar action that is not much (*kathīr*), the animal is lawful with that mention of Allāh's name. But if his conversation becomes prolonged or his action becomes much, eating it is disliked. In this there is no precise measure for when an action becomes "much," so the standard of the people is taken into consideration. If the people consider it to be much, it's much; if they consider it to be little, it's little. (Ibid. 5:288)

4. Ibn Qudāma states: "Mentioning Allāh's name over an animal to be slaughtered is considered correct if done at the moment of slaughtering or soon before, similar to when performing the ablution (*wuḍū'*).[42] Furthermore, if one mentions the name of Allāh over a sheep, but then takes another one and slaughters it with the first mention, it is not permissible to eat that animal, regardless of whether he lets the first one free or slaughters it. This is because he did not intend to slaughter the second sheep when mentioning Allāh's name. Likewise, if he sees a flock of sheep and says, 'In the name of Allāh' (*Bismillāh*), and then takes hold of a sheep and slaughters it without mentioning Allāh's

42 This is in reference to the Ḥanbalī school, in which mentioning Allāh's name is required immediately before performing ablution (*wuḍū'*). (Translator)

name a second time, it is not permissible to eat that animal. And if he was ignorant of the fact that his initial mention was not sufficient, he is not excused like one who forgets would be. Forgetting eliminates culpability for a mistake, but one who is ignorant is taken to task. It is for this reason that if one, out of ignorance, eats while fasting, his fast breaks, while this is not the case for one who eats because he forgets that he's fasting.

Moreover, if one lays down a sheep to slaughter it and mentions Allāh's name, but then discards the knife he is using and replaces it with another, or returns the greeting of a bystander, or speaks to someone, or takes a drink of water, or does something similar, and finally slaughters, it is lawful to eat the animal. This is because he mentioned the name of Allāh on that particular sheep and did not delay slaughtering except for a little bit, so it is as if he did not talk [or do any other action at all]." (*Al-Mughnī* 11:33)

5. Al-Mawwāq al-Mālikī states, "Imām Mālik said, 'Mentioning Allāh's name must be done when shooting an arrow or sending out hunting dogs or slaughtering manually, in accordance with Allāh's words "And mention the name of Allāh over it."'" (*Al-Tāj wa 'l-Iklīl bi Hāmish Mawāhib al-Jalīl* 3:219)

These juristic statements clearly show that the overwhelming majority of the Imāms who require one to mention Allāh's name when slaughtering also require that one do so over a particular animal. They also require that he do so just before he slaughters and that there not be a significant pause between mentioning and slaughtering. The fulfillment of these requirements is lacking in the automated system. Therefore, even if the one who turns on the machine mentions the name of Allāh when first starting it up, he has still not mentioned His name over any particular animal. On top of that, there is a long interval—perhaps several hours or an entire day or two—between his mention and the slaughtering of thousands of chickens. This one mention is obviously not sufficient for slaughtering all these animals.

This is similar to what Ibn Qudāma speaks of—one who sees a flock of sheep, then says, "In the name of Allāh," then takes hold of a sheep, and then slaughters it without a separate mention of Allāh's name. This is prohibited (*Al-Fatāwā al-Hindiyya* 5:289).

Some might find this argument dubious because of what some jurists have stated:

> If someone lays one of two sheep on top of the other, one mention of Allāh's name is sufficient if he slaughters them both with one swipe of the blade. Conversely, if one gathers some birds in his hand and slaughters them all after mentioning Allāh's name, and then immediately slaughters another one without making another mention, the last bird is not lawful. However, if he passes the blade along all of them, they would all be lawful with one mention. (Ibid. 5:289)

One might suspect that the automated system described above resembles the case of one who slaughters two sheep or several birds at one time, for which one mention of Allāh's name is sufficient. However, the truth is that these two examples are irrelevant to our case because in the above excerpt, the slaughtering of the two sheep, and of the birds, occurred all at once, without any significant pause between mentioning and slaughtering. This excerpt also asserts that if one gathers some birds in his hand and slaughters them after mentioning Allāh's name, and then slaughters another one immediately afterwards without mentioning again, the last bird is not lawful. This is precisely because there is a pause (however short it may be) between slaughtering the last bird and slaughtering the first birds, which were all killed at one time. Thus, in our case, we cannot say that all of the chickens slaughtered over the course of an entire day or two are killed all at once. Rather, there are many separate groups that are slaughtered one after the next. The difference is obvious.

This confirms that one mention of Allāh's name from whoever turns the machine on is not enough for slaughtering all of the chickens. And even if a worker were stationed at the rotating knife in order to mention the name of Allāh when each batch of chickens passes for slaughtering—this is something I saw done in a slaughterhouse in Canada—there are three complications that make this problematic.

First, the mention must come from the slaughterer, but the man standing in front of the rotating knife has no connection with the actual slaughtering process because he did not start up the machine, rotate the knife, or bring the chickens to the blade. He is completely detached from the process. In this case, therefore, a mention from him is not regarded as that of the slaughterer.

Second, a large number of chickens comes to the rotating knife in a matter of a few seconds. It is virtually impossible for the stationed man to mention the name of Allāh over every one of the chickens without any pause.

Third, the stationed man is a human being, not a machine. He is not able to completely avoid becoming occupied with anything besides mentioning Allāh's name on the chickens. It is more than likely that some need—a call of nature, for example—distract him from mentioning. Meanwhile, tens or hundreds of chickens may pass along the knife, being slaughtered without any mention at all. In the slaughterhouse in Canada, I myself saw this worker leaving his position for periods of time as long as half an hour or more.

There is another important point regarding the topic of mentioning Allāh's name over this automated machine. Suppose we were to analogize turning on the machine to sending out a hunting dog, for which it is not necessary to mention Allāh's name at the moment the game is killed but only when sending out the dog. It is allowed that there be a long interval between sending and killing, and it is also allowed that the dog kill a number of animals in one mission. Nevertheless, just one mention of Allāh's name is sufficient to make all of the animals lawful. Ibn Qudāma states:

> If a hunter mentions Allāh's name over [a hunting dog with the intention of catching] a particular prey, but the dog then catches another animal, it is lawful. But if he mentions Allāh's name over an arrow, but

> then discards it and replaces it with another and shoots it, whatever he strikes is not lawful. This is because when it is not feasible to mention Allāh's name on the game itself, the hunting tool is taken into account, unlike in [manual] slaughtering. It is conceivable, however, that one find this hunted animal to be lawful by drawing an analogy with the case of one who mentions Allāh's name over a knife and then discards it and uses another instead. But this analogy is invalid, because the relaxed obligation of mentioning Allāh's name over a particular hunted animal is owing to the exceeding difficulty present in doing so, and thus does not, as a logical consequence, translate into a similar relaxed obligation when voluntarily slaughtering an animal. (*Al-Mughnī* 11:33–34)

This statement pertains to forced slaughter (*dhakāt iḍṭirāriyya*), while our case pertains to voluntary slaughter (*dhakāt ikhtiyāriyya*), and the rules for the latter cannot be taken analogically from the former.

But when we take a look at the need for huge output in the least amount of time (owing to the increasing Muslim population), the growth in the number of consumers, and the seeming dearth of slaughterers to meet the high demand; and at the fact that because of the difficulty of mentioning Allāh's name while hunting, Islamic law has relaxed the requirement of doing so directly over the hunted animal, as Ibn Qudāma states; and at Islamic law's known practice of removing hardship in such situations—all this might seem to justify analogizing voluntary slaughter to forced slaughter for the ruling of mentioning Allāh's name (and for this ruling only), in order to remove difficulty from the matter and make it easy for people.

I myself am somewhat unconvinced of the strength of this argument, but I would like to present it before some scholars to research the matter and arrive at a proper and definitive resolution. As of yet, I have not issued a legal ruling in favor of this argument. This is mainly because there is a feasible alternative to the rotating knife that meets all production requirements in the same amount of time. That is that the rotating knife be removed from the process and in its place stand about four Muslim workers who take turns, as the hangers pass by, in cutting the necks of the chickens while mentioning the name of Allāh. I recommended this to a large slaughterhouse on Réunion Island. They put it into practice, and the results of the experiment showed that this technique does not decrease the volume of output at all, as these workers are able to cut the chickens' throats in the same amount of time as the rotating knife. Furthermore, this automated machine does not replace manpower entirely anyway. I have observed that factories are still forced to assign workers to stand in some of the areas where the hangers pass by, and these workers use their hands or hand tools to remove the chickens' innards. I am not aware of any slaughterhouse that is able to do without any manpower at all. So if these slaughterhouses station people for such purposes, they can also station four people at the slaughtering stage so that the slaughtering occurs by the Islamically prescribed method: at the hands of Muslim slaughterers who mention the name of Allāh when killing the animals. The machine can then carry out the remaining phases of the process.

In addition to Réunion Island, I saw the same alternative process put into practice in a larger slaughterhouse near Durban, South Africa, where output reached into thousands of chickens per day. They accepted the local Muslims' proposal to implement it and now use it without any problem.

I also spoke with the managers of the slaughterhouse I visited in Canada. I made the same recommendation to them, and they showed their readiness to comply if the Muslim residents requested it. Most unfortunately, however, the Muslim organization that issued statements certifying their meat as *ḥalāl* did not accept the proposal.

As long as this alternative is available, there is no great need to use the rotating knife or to apply the rules of forced slaughter to voluntary slaughter. And Allāh Most High knows best.

Chickens' Passage Through the Hot Water

There is one final issue regarding this automated system. And that is that the chickens, after leaving the rotating knife, pass through an area into which hot water pours down on them from above in order to remove their feathers. This hot water could cause two problems.

First, if the throats of some of the chickens are not properly cut by the rotating knife, those animals remain alive until they are immersed in the hot water. It is not unlikely that they die because of the water's intense heat, in which case they would become unlawful.

Second, it is sometimes uncertain whether immersion into the hot water occurs before the impurities (particularly blood) leave the inside of the chickens. It is quite possible, then, that these impurities permeate their meat because of the water's high temperature. The jurists have mentioned that such animals are never lawful. The author of *Al-Durr al-Mukhtār* writes:

> And likewise for a chicken tossed in hot water right after it has been slaughtered in order to be de-feathered before being cut into pieces.

Ibn ʿĀbidīn (may Allāh have mercy on him) explains this in his *Radd al-Muḥtār*, a commentary on *Al-Durr al-Mukhtār:*

> [Imām Ibn al-Humām] says in his *Fatḥ al-Qadīr,* "Such animals are never pure, but according to Abū Yūsuf they are. The reason they are not pure—and Allāh knows best—is that they absorb the impurities into their meat because of the boiling temperature of the water." (*Radd al-Muḥtār* 1:334)

This problem, however, is not an issue in our case because the water temperature does not reach boiling point; in fact, it is much less than 100°C (212°F). In addition, the chickens do not remain in the hot water for more than a few minutes, which is not enough time for the meat to absorb the impurities. The jurists who hold the view that chickens placed in hot water are unlawful only maintain this opinion if the water reaches boiling point and the chickens remain in it long enough for their meat to absorb their bodies' impurities. Ibn ʿĀbidīn further remarks after his above explanation:

> On a related note, it has become known that the *samīṭ* meat [that of an

> animal dipped in hot water to remove its fur, feathers, etc.] found in Egypt is impure. However, the aforementioned cause for the meat to become impure does not hold as long as the meat does not remain in the water, after it begins boiling, long enough for there to occur absorption and permeation of the impurities into the meat. Neither of these occurs in the *samīṭ* meat, since the water does not reach boiling point and the meat is only left in it long enough for the heat to reach the outside of the skin and the hair pores to release the fur. In fact, if it were left in there any longer, extraction of the hair would not be possible. (*Radd al-Muḥtār* 1:334)

This completely corresponds to the chickens' passage through the hot water in the automated machine. I have placed my hands in the water, and it was not burning hot, let alone boiling.

Conclusions Regarding the Automated Slaughter of Chickens

From the facts we have covered, we gather that this system for slaughtering chickens is deficient from the perspective of Islamic law in the following ways:

1. In some slaughterhouses, the chickens, before being slaughtered, are immersed in cold water containing a current of electricity. As such, it is feared that this current might cause them to die before they are slaughtered. Some experts have estimated that this electrical current causes cardiac arrest in ninety percent of the chickens. And Allāh knows best.

2. Even though the rotating knife does cut the jugular veins most of the time, some of the chickens' necks do not reach and make full contact with the blade. In such cases, it does not cut their necks at all, or only cuts them a little bit, and so their jugular veins remain uncut.

3. With the rotating knife it is not possible to mention the name of Allāh over every chicken. And a mention made when turning on the

machine, or by a person standing next to the knife, does not fulfill the requirements of Islamic law.

4. It is feared that the hot water through which the chickens pass might cause death in the chickens whose necks are totally uncut or only partially cut.

Analyzing the causes of these flaws, we see that they are not difficult to rectify. It is perfectly practicable to use the automated system for slaughtering chickens after implementing several modifications:

1. Do not use the electric current in the cold water, or verify with total certainty that it does not cause cardiac arrest in the chickens.

2. Do not use the rotating knife. Instead, station several Muslims or Kitābīs to take turns slaughtering the chickens—with their hands and with a mention of Allāh's name over every chicken—as they pass in front of them. I described this process in detail and noted that it has been implemented in several large slaughterhouses at the request of the local Muslims. It does not decrease the amount of output at all.

3. Be sure that the hot water through which the slaughtered chickens pass does not reach boiling point.

With the observance of these three adjustments, the chickens prepared by this machine will be completely lawful.

Industrial Slaughter of Other Livestock

The method for slaughtering livestock other than chickens, such as cows and goats, is different. They are not killed with an automated knife; people are involved in the process, which has varying techniques from slaughterhouse to slaughterhouse. One of these techniques is suffocation, as is popular in the so-called English Method. In it, the chest of the animal is pierced in the middle of the ribcage, and the animal is then inflated with air until it suffocates from the immense air pressure against its lungs. No blood exits the animal's body. It goes without saying that an animal slaughtered in this manner falls under the category of "what has been strangled" (*munkhaniqa*), which the Holy Qur'ān has explicitly prohibited. Earlier we concluded that it is unlawful to strangle (or suffocate) an animal, whether a Muslim or Kitābī does it. There is no way to make such an animal lawful.

The majority of today's slaughterhouses, however, kill these animals by causing some blood flow with a cut in the side or back of the neck. Since the methods of wounding the animal vary, we cannot be sure whether the jugular veins are cut or whether the animal is cut in some other place. An animal is not lawful until it is confirmed that its neck has been cut in accordance with Islamic law. If the slaughterer is Muslim, however, he is given the benefit of the doubt that he has slaughtered the animal properly—that is, that he cut the jugular veins.

For us, the main issue regarding the animals killed in these slaughterhouses is that the ownership insists on stunning or drugging the animals before someone actually slaughters them. In their view, stunning is necessary to give relief to the animals and lessen their pain when they are killed. To contain the animals, they use a number of devices designed to keep the animals from scrambling away and to hold their necks in place for easy slaughtering.

Stunning is done in different ways. Probably the most prevalent one is with a gun. This gun does not shoot lead bullets, but instead discharges a needle or a metal rod. They place the gun against the middle of the animal's forehead and shoot. The needle or rod bores into the animal's head, rendering it unconscious immediately. After that, they slaughter it.

Another way is with a large hammer used to strike the animal's forehead. This has proven to be painful for the animal, so many slaughterhouses have ceased using it and replaced it with the gun method.

A third way is the use of gas. They lock the animal in a room containing high levels of carbon dioxide gas, which affects its brain and leaves it unconscious. They then slaughter the animal by hand.

A fourth way is the use of electric shock. A tong-like device is placed on the animal's temples, and an electric current is sent through, zapping it and knocking it unconscious.

The juristic ruling for such stunning techniques requires analysis from two angles: (1) whether using such methods is itself permissible under Islamic law, and (2) whether the animal is lawful if a Muslim or Kitābī slaughters it with the Islamic method after it has been stunned.

As for the methods themselves, the permissibility of using them hinges on whether or not they really lessen the animal's pain when slaughtered. The Messenger of Allāh ﷺ, in a well-known ḥadīth, commands us to slaughter animals swiftly and to be gentle with them:

> When you kill, do it properly. And when you slaughter, do it properly; let each of you sharpen his blade, and let him put the animal he slaughters at ease. (*Ṣaḥīḥ Muslim*)[43]

Undoubtedly, the method that Islam has prescribed for slaughtering an animal—cutting the jugular veins in the neck—is the best way to kill it and the easiest for it to bear. Stunning it, on the other hand, in some instances harms the animal and pains it more than the actual slaughtering. The technique of beating it with a hammer on its forehead is, without a doubt, impermissible in Islamic law. As for the other methods, we cannot really be sure whether they decrease the animal's pain, as some claim, or actually increase it. Shooting a gun at its forehead involves nothing less than a violent blow; the electric shock cannot be without pain; and locking the animal in a gas-filled room causes breathing difficulty. Nevertheless, some experts insist that this does reduce the animal's pain. If it is shown conclusively that this is indeed

43 *Kitāb al-Ṣayd, Bāb al-Amr bi Iḥsān al-Dhabḥ wa 'l-Qatl.* This ḥadīth is also narrated in *Sunan al-Tirmidhī, Sunan Abī Dāwūd, Sunan al-Nasā'ī, and Jāmiʿ al-Uṣūl.*

the case, and that the animal does not die from being stunned, then it is permissible to use these methods. Otherwise, it is not.[44]

As for the animals slaughtered after being stunned, the correct ruling depends on whether or not the stunning causes them to die. Some experts assert that it does not cause death but only makes the animal lose consciousness and its ability to feel pain. However, this claim is debatable. Stunning with the gun violently strikes the animal's forehead, so it is not unlikely at all that it would die from severe brain damage, in which case it would fall under the category of "what has been beaten to death." I saw this technique used in Detroit, Michigan. The rod left the gun and penetrated the head of a cow to a depth about equal to a finger's length. Blood gushed out of its head, it instantly collapsed on the ground, and its body went totally limp as if it were dead. But the American owner of the slaughterhouse assured me that the animal remains alive for a few minutes after being shot, and that if it is not slaughtered within twelve minutes, it dies. I was not able to verify the accuracy of his claim. What I saw, though, made me doubt the assertion that stunning the animal in such a manner does not kill it. And from what I observed, there was nothing to remove the possibility that at least some of the animals do die from such a fierce blow.

In the case of electric shock, some experts have admitted that it can, in some instances, stop the heart. They also admitted that when the carbon dioxide gas reaches a certain level, it could possibly cause death.

This topic requires an in-depth, technical study from Muslim experts who are observant of and eager about their religion. Seeing as the subject is outside my expertise, I cannot be sure enough to make any definitive statements. I therefore recommend to the International Islamic Fiqh Academy, Jeddah, that it form a committee of Muslim experts to draw up and present a report after a thorough investigation of the matter. There is no doubt that if these stunning methods cause

44 Refer to the article in Appendix D entitled "Is Islamic Slaughtering Cruel to Animals?" There you can read about a study that confirms that the Islamic slaughtering method is the least painful to an animal, as well as the most beneficial to the consumer. (Translator)

death, or if it is feared that they cause death, then it is not permissible to use them, nor is any animal slaughtered after being stunned in such a manner lawful. As long as these methods are doubtful, the safest course of action is to avoid them completely. It is known that the Jews do not accept any form of stunning done to their animals. It behooves Muslims even more to stand clear of doubtful situations. And Allāh Most High knows best.

Appendix A

Survey of Meat Imported into Muslim Countries

[Author's Overview of the Topic]

THE MARKETS IN the Muslim world have become saturated with meat imported from foreign countries such as England, the United States, Holland, Australia, and Brazil. The juristic proofs presented in this book have established that animals slaughtered by the People of the Book are lawful for Muslims only if the slaughterers comply with the Islamic requirements for animal slaughter. The People of the Book were known to do this when Allāh revealed to His Messenger ﷺ the Qur'ānic verses that permit Muslims to eat their slaughtered animals. Today, however, the circumstances have changed quite a bit.

It is well-known that the Jews generally do continue to adhere to their religion's rules for slaughtering animals. They have organized private slaughterhouses under the supervision of their scholars and rabbis. Their meat is marked "Kosher" and is available almost anywhere Jews reside.

Christians, on the other hand, have completely divested themselves of the responsibility of slaughtering animals according to their sacred law. We have discussed a few passages from their scriptures that delineate some of the Christian laws of animal slaughter. They no longer adhere to these laws, which means that any particular meat slaughtered by a Christian is not lawful unless it is first verified that the slaughterer duly fulfilled the requirements prescribed by Christian law. Consequently, unverified meat sold in Western countries and imported into Muslim countries is prohibited. The grounds for this are many:

1. There is no way to know the slaughterer's religion with certainty. In the Western countries, there are polytheists, atheists, and agnostics in abundance, so there is no guarantee that any particular piece of meat has been slaughtered by a member of the People of the Book.

2. Even if we somehow knew for sure that the slaughterer is a Christian, we still would not know whether he is actually a Christian or espouses beliefs that negate the principles of Christianity. We have noted that a great number of Christians do not truly believe that a creator of the universe exists. Such people, in reality, are not Christians.

3. Furthermore, even if we somehow confirmed, or judged by outward appearance, that the slaughterer is a true Christian in his beliefs, it is still a fact that Christians today do not adhere to their prescribed laws of slaughtering. Some kill animals by strangulation, some kill them without cutting the jugular veins, and some use the questionable stunning techniques that we have discussed.

4. About the only thing that we can be sure of is that Christians today do not mention the name of Allāh when slaughtering. The dominant opinion upheld by the general body of the scholars and proven by the Qur'ān and Sunna is that mentioning Allāh's name is an invariable requirement for the permissibility of meat slaughtered by the People of the Book, just as it is for meat slaughtered by Muslims.

In view of these compelling reasons, it is not permissible for a Muslim to eat any of the meat sold in the Western markets until he is certain that whatever particular meat he is consuming at a given time has been slaughtered with a lawful method. Through the ḥadīth of ʿAdī ibn Ḥātim ﷺ, we have demonstrated the juristic principle that the meat of an animal is, by default, prohibited until proven lawful—that is, until it is verified that the animal has been slaughtered with a method acceptable under Islamic law. For this reason, the Messenger of Allāh ﷺ prohibited game for which stray dogs have competed in hunting down. Regarding hunting, the Messenger of Allāh ﷺ also said:

> If you find it drowned in water, do not eat it because you do not know whether the water or your arrow killed it. (*Ṣaḥīḥ Muslim*)[45]

Here again we see that when there is evidence that an animal has been killed in accordance with Islamic law and there is evidence to the contrary, preference is given to the latter, thus reinforcing the principle that an animal is prohibited until proven lawful. This is a key principle that many jurists have cited for this issue.

The same is the ruling for imported meat; the four reasons stated above apply here as well. A major problem today is that written statements placed on packaging material and shipping containers declare that the enclosed meat has been slaughtered according to the Islamic method. To address this problem, the Board of Senior Scholars in Saudi Arabia (*Hay'at Kibār al-ʿUlamā'*) commendably undertook the task of dispatching representatives to the different foreign slaughterhouses that export meat to Muslim countries. These representatives then sent their reports of what they saw in those slaughterhouses, and these reports clearly demonstrate that the claim written on the shipping containers is not credible. A look at some of these reports should be sufficient to confirm our conclusion about meat imported from non-Muslim countries.

---

45 *Kitāb al-Ṣayd*. For further explanation of this ḥadīth, refer to *Takmila Fatḥ al-Mulhim* 3:494.

Report Summaries

Meat Imported into Saudi Arabia from Brazil, Argentina, & Denmark

Professor Aḥmad Maḥāyirī

Professor Maḥāyirī, in 1398 AH (1978 CE), traveled to seven Brazilian cities: Curitiba, Ponta Grossa, Campo Grande, Cuiaba, Goianá, Presidente Prudente, and San José.

Baransisa Company for Chickens (Ponta Grossa, Brazil)

- Slaughters over 150 tons of chicken meat per month.
- At the request of importing Muslim countries, the company employed the services of Ḥusayn al-ʿUmayrī (president of the Islamic organization in nearby Curitiba), for a commission of ten percent of the shipping cost, to sign a written statement declaring that the enclosed meat had been slaughtered according to the Islamic method.
- Problems of slaughtering method:
  - The chickens are hung upside down on a machine that takes them to a man who cuts their neck so hastily that one or both of the jugular veins may go uncut.
  - The chickens are very soon after placed in hot water for feather removal; those with uncut jugular veins may die from the hot water rather than the knife incision.
  - It is unclear whether the slaughterer is a Muslim or Kitābī or whether he is of some other creed.
- The professor called a meeting with the company officials and requested that the company modify the slaughtering method so that it would be in accordance with Islamic law. The manager agreed with the condition that the required amount be specified to the company in advance.

Sadia Oeste

- One of the largest meat corporations in the world. Monthly export volume for chicken meat alone reaches about 300 tons. It receives *ḥalāl* certification statements from Sao Paulo (which is about 1,800 kilometers away).
- The method of slaughtering chickens at this company is similar to the one at Baransisa.
- Method of slaughtering cattle:
    - The company manager mentioned that they formerly shocked the animals with electricity and then skinned the animals without letting any blood leave their bodies, but stopped doing so when they saw that the meat quickly spoiled and took on a dingy gray tinge. Veterinarians directed them to kill the animals such that the blood leaves the body. The manager said that they now slaughter the animals by cutting their jugular veins.
    - The professor reported seeing the following: the animal is knocked unconscious with a hammer blow to the head; the animal is then hoisted in the air, and a man with a knife slits its neck and cuts into its jugular veins, causing blood to gush out profusely until the animal dies.
    - It is unclear what the slaughterer's religion is, and the credibility of the *ḥalāl* certificates is seriously in question since the Muslim organizations that issue them are over 1,800 kilometers away.

Argentina

- At a cattle slaughterhouse, a machine lifts the cattle in the air, and a man cuts the jugular veins of each animal separately—that is, in accordance with the Islamic method.
- However, it is not clear if the slaughterer is a Kitābī or something else.
- A nearby Islamic center issues a written certificate of approval for each shipment.

Denmark

- Muḥammad Abyaḍ al-Maghribī, a worker at a meat-packing factory, said the following regarding meat exported to Arab countries: "They write 'Slaughtered According to the Islamic Method' on the packaging, but this is not true because the slaughtering here is always done using electrical shock."

Meat Imported from London and France

Shaykh ʿAbdullāh ibn ʿAlī al-Ghaḍiyya

- The shaykh visited London and asked for some information from the manager of a company that sells slaughterhouse machinery. The manager gave him an illustrated catalogue and explained to him the slaughtering process generally used in England for chickens:

    - The chickens are delivered in droves to the factory, where they are hung upside down by their legs. Each one's head is put into a machine displaying a sign saying "Slaughtering by Stunning." This machine subjects the chickens to a violent blast of air, which renders them blind and deaf. Whatever little bit of blood and other fluids that exits their bodies is collected in a nearby container. They are then taken through another machine that exposes them to extremely hot steam or scalding hot water in order to kill any chickens that are still alive. The meat is then cleaned and packaged in nylon sacks on which is written "Slaughtered According to the Islamic Method."

- The manager said that the same technique is used in France.
- Other notes from the shaykh:

    - The heads of chickens slaughtered in London and other places in England are removed only when shipped to the Middle East.

- Chickens with heads intact and without any visible knife-marks on their necks are readily available in cities in England.
- The animals' blood is not allowed to drain out, in order to give the animals added weight and boost the company earnings.

Meat in Greece

Professor Jamāl ibn Idrīs al-Yūnānī

- The professor writes to Shaykh ʿAbd al-ʿAzīz ibn Bāz about the slaughtering methods used in Greece.
- Small animals are slaughtered as required by Islam; that is, their blood is allowed to exit their bodies and only then are they skinned and cut into pieces in preparation for packaging and shipment.
- Large animals, however, are shot in the head with a projectile fired from a gun-like implement and then slaughtered before they die, but it is uncertain whether they remain alive until they are slaughtered or whether they die beforehand.
- As an aside, the professor informs us that according to a veterinarian whom he met, the color of the bones of a slaughtered chicken is a very good indicator of how the animal was killed. If the color leans more to white, it is a good sign that the animal's blood was allowed to drain out through a proper Islamic slaughter. If the bones have a more blackish tinge, however, there is a good chance that the animal was not slaughtered according to the Islamic method.

Methods Used in British Slaughterhouses

Shaykh Ḥasan ʿAbd al-Ghaffār

- Someone requested the shaykh to give a description of the slaughtering methods used in British slaughterhouses that export meat to

Muslim countries in the Middle East; he also requested an overview of the ill nutritional effects of slaughtering using such methods.

- The shaykh's answer regarding sheep and cows:
  - Shocked with electricity.
  - Muslim butcher is hired to Islamically slaughter a small number of animals reserved for local consumption.
  - Non-Muslim butcher slaughters the rest by randomly thrusting the knife into the animal's neck and allowing some blood to flow out.
- The shaykh's answer regarding chickens:
  - Stunned with electrical current.
  - Necks then automatically cut with a sharp knife and blood allowed to exit their bodies.
- The ill effects according to a study conducted in the UK by a number of Muslim doctors and scientists:
  - Stunning the animal before slaughtering it creates weakness in it and causes its heart to constrict. Thus, a large amount of blood does not leave the body, resulting in poor nutritional value and unpleasant taste.
  - If the electric shock is light, the animal feels pain from it as well as from the actual slaughtering. And if it is intense, it stops the animal's heart, thus killing it before it is properly slaughtered.
  - The claim that stunning puts the animal at ease is not correct. The purpose is only to slaughter as much as possible in the least amount of time in order to increase company profits.

Meat in Yugoslavia

Shaykh ʿAbd al-Qādir al-Arnā'ut.

- In the small villages, Muslims hand slaughter goats according to the Islamic method.
- In Sarajevo, the capital of Bosnia-Herzegovina, Muslims give a small shock to their goats in order to make them fall to the ground. While the animals are still alive, they are placed under a guillotine-shaped implement that cuts their heads off and causes some blood to flow out. The animals are then processed and packaged, and "Sarajevo" is written on the packaging.
- In other cities, the slaughterer may very likely be a non-Muslim, perhaps a Christian or Jew or perhaps a Communist.
- The exported meat comes from animals that are hand slaughtered by Muslims, as in the small villages.
- In a number of venues, there is a high possibility that the knife used to slaughter goats may also be used to slaughter pork.
- It is recommended not to purchase any meat slaughtered in the city of Sarajevo.

Findings of a Study by Professor ʿAbdullāh ʿAlī Ḥusayn

- The professor is a scholar from Al-Azhar, Egypt. In his book *Al-Luḥūm: Abḥāth fī al-Dhabā'iḥ wa 'l-Ṣayd wa 'l-Luḥūm al-Maḥfūẓa* (*Study on Slaughtered Meat, Hunting, and Canned Meat*), the professor concludes that canned meats imported into Egypt from Europe, Australia, and the United States are completely unlawful, as the animals are bludgeoned to death.
- To get an official report of the slaughtering methods used in various exporting countries, he wrote to authorities in fourteen countries asking three questions: (1) What slaughtering methods are used in your country? (2) What part of the animal's body is struck/cut first when it

is killed? (3) What canned meat products are produced and exported from your country? He received responses from Turkey, Greece, Holland, Spain, and Denmark. The most glaring examples of failure to meet the Sharīʿa's requirements were in the responses from Holland and Denmark.

> *Method in Holland.* Animals are stunned with a device that knocks them unconscious instantly. As cutting the head and neck with a knife is prohibited by law, the animals are killed with a helmet-like implement that shoots a rod into their heads.
>
> *Method in Denmark.* Larger animals (such as bulls and cows) are killed with a projectile—either special bullets or large nails—that is shot into the head. Smaller animals (such as goats) are killed with a bullet or a hammer-blow to the head. Poultry is killed with either a hammer-blow to the head or a severing of the head altogether. After slaughtering, the blood of the large and small animals is released by cutting into the aorta at the top of the chest with a regular knife.

- The professor then comments: "All of this information is official, and it conclusively proves our assertion that their slaughtered animals are beaten to death and are thus impure and unlawful. No Muslim is allowed to consume, carry, or sell them. During my five years as a student in Europe, I was personally convinced from the information I had that the method of animal slaughter used there was to give the animal one blow to the head—using some specialized device—so forceful that it caused the animal to collapse immediately and die. However, afraid that this might have been an untrue claim, I obtained written proof from the governments of the exporting countries. And here I have presented the facts so that people may know the truth once and for all."

- The professor then says regarding American meat: "I later sent a message to the respected professor and shaykh Dr. ʿAbd al-Ḥamīd Muṣṭafā Farghalī, a specialist in animal organs at Johns Hopkins University, Baltimore, Maryland. I asked him about the method of animal slaughter used in the United States. He replied . . . saying, 'I asked about the slaughtering method. What they do is strike the animal on its head with a sharp hammer, killing it. After that, they cut into its neck. They do not slaughter as Muslims or Jews do, and they use this method for all types of animals.'"

Articles

"Let Us Not Eat What Is Ḥarām"

Dr. Maḥmūd Ṭabbāᶜ

Al-Daᶜwa Magazine, Dhū 'l-Qaᶜda 21, 1398 AH (1978 CE)

Assalāmu ᶜalaykum wa raḥmatu 'Llāhi wa barakātuh

I recently came across an article in *Al-Daᶜwa* magazine also entitled "Let Us Not Eat What Is *Ḥarām,*" written by ᶜAbd al-Raḥmān Muḥammad Ismāᶜīl (may Allāh reward him). I would like to add some information of my own.

My name is Dr. Maḥmūd Ṭabbāᶜ, and I am a veterinarian. I did my studies in West Germany. At the beginning of my studies, I, along with my Muslim brothers, faced the problem of whether the slaughtered meat there was permissible to eat or not. To check the slaughtering method, I went with several of my brothers to visit a slaughterhouse in Hannover. There we saw butchers taking a group of cows and shooting each one in the head with a special gun. When the whole group of animals lay motionless on the ground, the workers took a twenty-minute food break. After eating, they began hanging the animals by their hind legs on a moving hoist and then proceeded to cut off their heads. They then stripped each animal's skin, split its entire body in half, and cleaned it out with water before removing its innards. The water they used to wash the animals turned blood red.

We were totally positive that the cows were already dead before their heads were severed. Thus they clearly fell in the category of carrion (*mayta*), which Allāh has prohibited to us Muslims. We made every effort to inform the Muslims students in our area of what we saw and advised them not to eat such meat. But most of them did not even hesitate to eat pork, so what could have possibly stopped them from eating this meat?

"The Legitimacy of Slaughtered Chickens in Denmark"

by the Muslim Youth Group of Denmark

Al-Mujtamaᶜ Magazine, Dhū 'l-Qaᶜda 1, 1398 AH (1978 CE)

After receiving much correspondence from Muslims residents in Arab countries requesting accurate information about the slaughtering method used for poultry and other meats exported from Denmark, we have conducted extensive research and investigation and have arrived at a number of important results. Here we present our findings.

We have learned from official sources that the Qadianis, ever since they established themselves in Denmark in 1967 CE, have been acting on behalf of the Muslims in Denmark. They have been issuing *ḥalāl* certification for shipments of poultry and other meats bound for Muslim countries and in return receiving remuneration from the exporting firms. We have also come to know quite well that the embassies of Muslim countries found here in Denmark, very much like those in other countries of the world, do not represent the true spirit of Islam whatsoever. They conduct their dealings with their own self-interests in mind, holding diplomatic gatherings and pandering to the wishes of the leaders of Muslim countries, in hopes of receiving promotions and lining their pockets handsomely. However, lest we make any blanket statements, we must note that there are several exceptions—a number of individual Muslim workers here and there whom Allāh has protected from placing professional ambitions above obedience to Him. These people, nevertheless, are in the minority.

Furthermore, over the past few years we have discovered that some meat companies resort to tricky tactics to sell Danish-slaughtered chickens to Muslims countries. One such tactic is playing tape recordings of Qur'ānic recitation inside the slaughterhouse, as if this would somehow make the meat lawful for Muslims to eat. To further cover up their activities, some slaughterhouses station one or more Muslim workers in certain conspicuous parts of the factory to do small tasks that have no connection with the actual slaughtering. And even if one of them did do some slaughtering, it is not conceivable that he would be able to slaughter all of the thousands of chickens that the slaughterhouse produces daily. Libya was one of the first Muslim countries to detect this farce and resolved at once to prohibit by law the importation of chickens and other meat from Europe. Allāh knows best whether this restriction is still in effect.

Over the last ten years or so, the problem of chickens imported from Europe hardly concerned most Muslims as this issue was overlooked in the midst of political turmoil that was, and still is, chafing away at Muslim unity in the Middle East. Some people took advantage of the situation by circulating Danish-slaughtered meat in the Middle East under the pretext that the people in Denmark were of the People of the Book. We firmly reject this assertion. It takes no more than a brief glance at the societal evils in Denmark to see that the people there without a doubt cannot legitimately be deemed among the People of the Book. Fornication and adultery, nudity, alcohol, gambling, homosexuality, severance of family ties, disobedience to parents, interest, and many other evils have become accepted and are raging rampantly throughout the country. There is even official national legislation protecting people's rights to perform such horrendous acts. The "Christians" in Denmark are more Communists than they are Christians.

In a natural reaction to the suspicious activities of some slaughterhouses, our youth group felt the burden of responsibility to investigate the issue by contacting all of the Denmark slaughterhouses (thirty-five in total) that shipped chickens abroad. Our correspondence to each slaughterhouse ran as follows:

> In the past months, we, an educational Islamic organization based in Denmark, have received inquiries from Muslims living inside and outside Denmark regarding the methods used to slaughter chickens and other poultry that are exported to Arab countries. An adequate answer to these question is extremely important to Muslims, as the slaughtering method must be in accordance with the laws laid out in the Holy Qur'ān. For this reason, we request that you allow a group of three to four people from our organization to visit your slaughterhouse in order to inspect the slaughtering method used. We also request that you allow us later on to publish the findings of the inspection in the monthly Arabic magazine *Al-Ṣirāṭ,* so that Muslims may be better informed about the issue. And, of course, we will be sure not to speak of your company in an ill manner. We look forward to your swift response.

Upon receiving the responses, we found out that some of the slaughterhouses to which we wrote in fact did not export their meat to Muslim countries at all; naturally, these slaughterhouses did not resist our visiting them. However, those slaughterhouses that did export to Muslim countries, in one way or another, did not grant us permission to visit them. Some of them outright refused to allow us to come; others simply referred us to a committee of professionals in charge of poultry exportation, telling us that this committee represented their slaughterhouse and that it held the responsibility of investigating any issues revolving around Islamic slaughter. After trying endlessly to call this committee, we finally got through, but the committee refused to allow us to conduct any type of inspection, claiming that our organization did not represent Islam or Muslims in Denmark. They told us that they were in contact with another Muslim association in Denmark, which in their minds did represent Islam, as it was connected with a number of Arab foreign embassies. This Muslim association okayed the slaughtering method and issued *ḥalāl* certificates knowing full well that the exported chickens were in no way different from all the others, with the exception of the words "Slaughtered According to the Islamic Method" written on the packaging.

After investigating a bit more, we found that this association was not run by Qadianis, as had been the case over the preceding ten years. Instead, another Islamic organization had taken over the task of issuing *ḥalāl* certificates—as well as the baggage that comes with that task, such as the responsibility of defending the interests of the slaughterhouses and the interests of Arab importers and the embassies standing behind them.

From a phone conversation with the manager of this new organization, we learned the following:

1. The Danish slaughterhouses do not have the slightest idea about the requirements of Islamic slaughter. Whatever information they have is no more than a series of baseless claims from random conversations with groups of ill-informed Muslims. Many of these Muslims' statements contradict each other, so the end result is that the Danish slaughterhouses do not think that Islam has any precise juristic ruling in the matter of animal slaughter.

2. The Arab importers are the ones that request that "Islamically Slaughtered" [in Arabic] be written on the shipping containers; they even prepare the labels. The Danish slaughterhouses agree to do so, as long as it does not hinder their sales and the "official" Islamic organizations continue to give *ḥalāl* certification to their shipments.

3. Slaughtering consists of stunning and then decapitation, as dictated by Danish veterinarian law. Slaughtering any other way requires a special permit.

4. The chief concern of Danish slaughterhouses revolves around the embassies of the importing countries. These embassies approve the issuance of *ḥalāl* certificates by the Islamic organizations in Denmark that "overlook" the slaughtering process. As long as these organizations

and the foreign embassies see eye to eye, then in the eyes of the Danish slaughterhouses, it is in no one's interest to tip the balance.

At the end of our conversation, we requested the manager to send us a written explanation of the points we discussed, but he delayed in responding. We did eventually receive a letter from him, but it was nothing more than a diplomatic response that did not come close to addressing the points brought up over the phone.

Upon assessing the facts, it is fairly apparent that the main culprit in this whole fiasco is not the Danish exporting slaughterhouse, but the Arab importer (and the embassy that stands behind it and gives its stamp of approval). In response to this, we, the Muslim Youth Group of Denmark, wish to declare to all Muslims, wherever they may be, that the meat exported to Muslim countries from Denmark is no different from the meat exported to other countries. The slaughtering of all the animals is done by decapitating them after stunning them. The only difference is in the special packaging that bears an Arabic phrase ["Islamically Slaughtered" or the like] used to trick the Muslim consumers.

The main objective of this investigation of ours has not been to produce an Islamic solution to the problem of imported meat. Our goal, rather, has been to confirm the facts surrounding the faulty slaughtering techniques and report our findings to Muslims so that they may themselves collectively endeavor to arrive at a suitable solution. But before leaving you to embark on a search for the solution, we would like to make a couple final remarks that may be of use in this task.

Several years ago, some Muslim brothers negotiated with a Danish meat company to make a special arrangement whereby Islamically slaughtered meat would be shipped to Muslim countries. The company agreed on the condition that the brothers themselves provide Muslim workers who they could guarantee would stick to their jobs and not quit. While searching for people willing to take this job, it was further learned that the slaughterhouse was located in a small town far away from any of the major cities. Naturally, this did not excite too many people into accepting the job, as most workers (especially foreign immigrants, which is the case for many of the Muslims) would much rather live in the capital or at least a major city. Foreigners are usually more able to adapt to the social environment of a large city, and in general workers like to have ease of mobility and access to news and other resources, which are not easily available in remote villages.

Therefore, making lawful meat available to Muslims—in an acceptable manner, whereby both the demands of the Sharīᶜa and the needs of the people are met—requires sincerity and commitment on the part of a few Muslims. It also requires people with knowledge of the Sharīᶜa's rules regarding Islamic slaughter, as well as a suitable budget to meet the financial demands of the project. Our youth group is fully prepared to do its part, by making calls to the slaughtering-equipment companies, drawing up a contract with a consultant firm, and forming an estimate of the needs and costs of the project.

There remains just one final remark. It does not pertain to slaughtered meat per se, but it does pertain to all Muslims who eat it. It is clear that very few Muslims today fully look into whether what they consume is lawful. The majority do not even stop to think about it and sometimes think that doing some investigation is too much of a hassle. Unfortunately, this is one of the trials of the times we live in. For most, gratification of the basest human urges holds top priority. What's more, many people fall back on, and in the process abuse, the statement "Indeed, Allāh is all-forgiving, mercy-giving." Then they brazenly entertain high hopes of somehow entering Paradise and setting eyes on Allāh Almighty without having made even the slightest effort or sacrifice.

Indeed, there are many Muslims who will travel long distances—perhaps tens of miles at a time—to get their hands on some desirable piece of merchandise or delectable dish of food. When a physical obstacle comes in their way, such people overcome it, not for the sake of pleasing Allāh and His Messenger, but just to quench their thirst. When an Islamic law comes in their way, they play games with it and circumvent it in one way or another. They say, "*Dīn* is ease," but never do they say, "With hardship comes ease." The fact is that entry into Paradise is not possible except through bearing some hardships in this life.

A perfect example of people who endure the difficulty of looking into what they eat can be found in vegetarians, who avoid all meat and products containing meat-derivatives. These people are known around the world for their meticulousness in investigating their food. Some go so far as to not even eat cakes and jams out of fear that they might contain oils extracted from animals. They unfailingly ask about the ingredients of certain food items before buying them, and they have their own special stores pretty much everywhere.

These people have laid down their own laws by themselves, and they respect them and live up to them. To them, investigating their food is an obsession. So what is it with Muslims? The Book of Allāh and the Sunna of His Messenger come to them, yet why is it that they pay no attention?

We ask Allāh that He benefit us and the reader with what we have said, that He make it a means of salvation on the Day of Judgment, and that He make our actions sincerely for His sake alone.

Legal Ruling of the Board of Senior Scholars, Saudi Arabia

Issue 4: Sharīʿa's stance on imported meats in light of the known facts. Simply explaining the proper Islamic method of slaughtering, without issuing a ruling for meat imported into Saudi Arabia from Europe, America, and other countries, is of no benefit to one who endeavors to eat what is lawful and avoid what is not. The only way one can know for himself what is lawful is if he is knowledgeable of both the Islamic laws of animal slaughter and the slaughtering methods used by the foreign exporting slaughterhouses. For most this is very difficult, as traveling to those foreign countries is quite costly. Most who make the trip do so out of some necessity, such as medical treatment, or for tourism or to pursue some other personal desire of theirs. Few make the journey to find out the truth about the slaughtered meat that is exported to Muslim countries.

To obtain some facts, the General Council for Academic Research and Legal Rulings wrote to Saudi officials responsible for overseeing the importation of meats and other foods products. The Council advised them to be diligent in assuring that everything imported was lawful according to the Sharīʿa, in order that Muslims may have lawful foods to eat and avoid consuming what Allāh has prohibited.

A vague, unsatisfactory response came back. So the Council wrote to some brothers doing *daʿwa* in Europe and America and requested them to find out about the slaughtering methods used and the religious affiliation of the slaughterers. One group of brothers wrote back with a somewhat ambiguous response. Another group of eager Muslims published articles in magazines describing the details of the slaughtering process and the slaughterers. May Allāh bless all for their efforts.

However, aside from the vagueness of some of the information, these exchanges did not include information regarding the particular slaughterhouses from which meat was imported into Saudi Arabia. Nevertheless, based on the information that we do have—reports from official representatives, relevant magazine articles, and other legal rulings regarding slaughtered meat—we present our findings:

1. In a letter to the president of the General Council for Academic Research and Legal Rulings, the general secretary of the World Muslim League writes that he received reports regarding some of the Australian companies that exported meat to Muslim countries. In particular, he mentions that Al-Halal al-Sadiq, a slaughterhouse owned and operated by Qadianis, does not follow the Islamic method of slaughter for cows, goats, and poultry. Based on this information, it is prohibited to eat the slaughtered meat of this company.

2. Professor Aḥmad ibn Ṣāliḥ wrote a report on the slaughtering method at the Baransisa Company for Chickens. In it he informs that

the religion of the slaughterer is unknown and that there is doubt as to whether either of the external jugular veins of some of the animals is cut during the slaughtering process. Furthermore, the *ḥalāl* certification is granted by an agent who does not indicate whether he himself, or someone on his behalf, actually oversaw the slaughtering process. Based on this, it is not permissible to eat any meat from this company. Additional evidence that the slaughtering method at this company is not in accordance with Islamic law is that the company manager agreed to modify the method so that it would be Islamically acceptable (with the condition that the required amount of meat be indicated beforehand).

4. Shaykh ʿAbdullāh al-Ghaḍiyya sent us a report about the slaughtering method used in London. He mentions that the slaughterers are mostly young pagans and atheists. The chickens leave the slaughtering machine dead and de-feathered, with heads uncut and without even a single cut-mark on their necks; the slaughterhouse personnel confirmed this with a statement written in English. Furthermore, when a Muslim inquires about the automated slaughtering techniques used in the slaughterhouse, he is shown to a small section of the slaughterhouse where a few Muslims are found slaughtering animals for other Muslims—a seemingly tricky ploy that raises doubt about the normally used slaughtering method and the religion of most of the slaughterers in the slaughterhouse. In light of this information, it is not permissible to eat the meat coming out of London.

5. Professor Ḥāfiẓ sent a report illustrating the slaughtering method used in some well-known places in Greece. In it he informs that large animals are slaughtered after being knocked out with a gunshot blow to the head. There is obvious doubt as to whether the slaughtering takes place before the animals die as a result of the head shot. Based on this report, eating this meat is not permissible. There is another process about which the professor states, "The slaughtering is done according to the Islamic method." However, he does not mention the specifics of the

process or the religion of the slaughterers, nor does he mention where in Greece the slaughterhouses that use this method are located.

6. Shaykh ᶜAbd al-Qādir al-Arnā'uṭ sent a report about the slaughtering method used in Yugoslavia. In it he states that the slaughtering process used in small villages and in Sarajevo is in accordance with the Islamic method and that the slaughterers there are Muslim. Based on this information, it is permissible to eat this meat. However, he also mentions that in other cities, there is equal possibility that the slaughterers are Muslim, Kitābī, or Communist; it is not clear which one they are in reality. Because the religious affiliation of the slaughterers is doubtful, eating the meat from these cities is not permissible.

7. Dr. Ṭabbāᶜ sent a report about the slaughtering method used in West Germany. He described that the cows there are first shot in the head with a gun and are not "slaughtered" until they are already dead. This meat may not be eaten.

8. *Al-Mujtamaᶜ* magazine published an article in its 414th issue about the slaughtering process used in Denmark. It explains that the slaughterers are closer to Communism and paganism than they are to Christianity. It also mentions that the slaughtering companies there do not have any idea about the proper Islamic method of slaughtering—all they have is faulty information picked up from local Muslims here and there—so as to be able to implement a process that is compliant with Islamic law. In addition, the phrase "Slaughtered on the Islamic Method" is written on the shipping containers, but this is done by the importers (in the Muslim countries) in order to convince anyone who may be doubtful that the meat is in fact lawful. On top of that, these companies prohibit anyone who wishes to learn about the slaughtering process from entering the slaughterhouse and making their inquiry.

Furthermore, Professor Aḥmad Ṣāliḥ Maḥāyirī and Muḥammad Abyaḍ al-Maghribī, the latter of whom works in the Danish meatpacking industry, write that the animals were claimed to be slaughtered on the Islamic method. However, this is erroneous, they say, because the animals are in reality killed by electric shock in all instances. Based on the information presented by these two and by *Al-Mujtama*ʿ magazine, eating the meat coming out of Denmark is not permissible.

9. Ibn al-ʿArabī reports that it is permissible to eat the cattle, poultry, and other meats slaughtered by the People of the Book, even if their method is not in accordance with the Islamic rules of slaughtering. He also claims that everything they deem lawful in their religion is also lawful for us, except for what Allāh deems to be to the contrary. In light of the above overviews of the slaughtering methods used in various parts of the world, as well as the many legal rulings (*fatwās*) surrounding the issue of slaughtered meat, this opinion of his is duly refuted and, therefore, must be rejected.

10. In light of the reports detailing the slaughtering processes and religious affiliation of the slaughterers, it is evident that the letters of the Ministry of Trade and Industry to the Council do not inspire any confidence that the meat imported in Saudi Arabia is lawful for consumption. There is still legitimate doubt as to whether the slaughtering methods used in the exporting countries are in accordance with the prescribed Islamic method. And the default ruling is prohibition. With that said, it is essential to search for a feasible solution to this problem.

Issue 5: Proposed Solution to the Problem of Imported Meats

A possible solution is summed up in the following four points:

1. It is necessary to breed and raise more animals; to import the needed amount of *live* animals to Saudi Arabia; to make available the different feeds that the animals need; to prepare a suitable place in Saudi for raising and slaughtering them; and to help the people who undertake this task by establishing a company to support them and providing manpower to aid them, in order to encourage their efforts and make it easier to distribute the meat throughout Saudi. Additionally, cheese

and meatpacking factories must be established, as well as factories that produce butter, fats, cooking oil, and other oils.

2. It is necessary to establish special slaughterhouses for Muslims—in which the Islamic method of slaughter is strictly adhered to—in the countries from which Saudi Arabia and other Muslim countries import meat.

3. At these special slaughterhouses, it is necessary to select trustworthy Muslim employees who are thoroughly informed of the Islamic method of slaughter to properly slaughter animals in the amount needed for importation.

4. It is necessary to select trustworthy Muslims who are qualified experts in the Islamic laws of slaughter and the food industry to supervise the slaughtering process as well as the production of cheese and other food products exported to Saudi Arabia.

The Jews are eager to assure that their slaughtering processes are in perfect accordance with their beliefs and principles, to the extent that they have established their own slaughterhouses and hired their own workers who slaughter the animals as they wish. If this is the case with the Jews, then it is even more befitting of Muslims to do the same. Muslim consumption of meats and foods produced in the West is huge. The Muslims are in great need of producing their own meats and other food products.

ʿAbd al-ʿAzīz ibn ʿAbdullāh ibn Bāz (Chairman)
ʿAbd al-Razzāq ʿAfīfī (Vice Chairman)
ʿAbdullāh ibn Ghadyān (Board Member)
ʿAbdullāh ibn Qaʿūd (Board Member)

[Author's Final Remarks]

What we have shown above is sufficient evidence to substantiate that most of the written certificates alleging that the imported meat is slaughtered according to the prescribed Islamic method are completely unreliable. Therefore, it is not permissible to eat such meat until it is proved credibly that it has been slaughtered in the proper manner.

Furthermore, it may have been noticed that some of these documents have dates going back a decade or two or even more. However, to the best of our knowledge, neither has a proper survey been conducted after these reports were issued nor has any basic improvement been made in this regard. Until credible proof is shown that the imported meat is slaughtered in accordance with Islamic law, the findings presented above stand.

Appendix B

Summary of the Book

Main Points

1. Unlike mundane things like cooking methods, the matter of slaughtering is not an ordinary matter that is ungoverned by Islamic legal injunctions. Rather, it is a "ritual" (*taʿabbudī*) act—one that is subject to precise laws prescribed in the Qur'ān and Sunna. Moreover, it is one of the fundamental rites and defining aspects (*shaʿā'ir*) of Islam, one of the marks by which a Muslim is distinguished from people of all other creeds. Regarding this the Prophet ﷺ said, "Whoever prays as we [Muslims] pray and faces in the direction of the Kaʿba [during prayer] and eats our slaughtered animals, that is a Muslim for whom is the protection of Allāh and His Messenger" (*Ṣaḥīḥ al-Bukhārī*).

2. Even if it is not one of the categorically prohibited animals (such as swine), a creature's meat does not become lawful (*ḥalāl*) unless slaughtered in the prescribed manner, which has the following requirements:

   (a) The act of killing a tractable animal must be done by way of cutting into the jugular veins (*awdāj*) (with slight difference among the scholars regarding the least amount that must be cut).

(b) The slaughterer, in addition to being sane and of the age of discretion, must be a Muslim or a Kitābī (i.e., a Jew or Christian).

(c) The name of Allāh must be mentioned at the time of slaughtering. If the mentioning of Allāh's name (*tasmiya*) is abandoned intentionally, the slaughtered animal takes the same ruling as carrion (i.e., prohibition), according to the consensus of the jurists (*ijmāʿ al-fuqahāʾ*). This view is supported by clear and definitive (*qaṭʿī*) proof texts. As for one who forgets to mention Allāh's name, his mistake is overlooked, and his slaughtered animal is lawful for consumption.

Also, the statement ascribed to Imām al-Shāfiʿī—that the slaughtered meat of one who intentionally abandons the mention of Allāh's name is lawful—is not established by any reliable source. In fact, his very own words in *Kitāb al-Umm* confirm that he only advocated permissibility in the case of forgetting and clearly declares prohibited (*ḥarām*) an animal over which the mention of Allāh's name has been abandoned scornfully.

3. Meat slaughtered by the People of the Book (*Ahl al-Kitāb*) was only allowed because they adhered to their religiously prescribed rules of slaughter. They prohibited carrion, that which has been strangled, that which has been beaten to death, and the dead prey of predatory animals, as mentioned in their sacred books. A number of such passages have been quoted in this book. The People of the Book used to mention nothing but the name of Allāh. Because of all this, their slaughtered meat was considered like that of Muslims and was thus made lawful.

4. Likewise, the women of the People of the Book were made lawful for Muslim men to marry because they adhered to marriage laws similar to those of Islam. Thus, for a marriage with a Christian or Jewish woman to be lawful, it is obligatory that the marriage take place in accordance with Islam's prescribed laws. According to the consensus of the scholars, the verse saying that their women are permissible for Muslim men to

marry (Qur'ān 5:5) is subject to the adherence of the two marrying parties to Islamic marriage laws. By the same token, Allāh's words "And the food of those who have been given the Book is permitted to you" (also in verse 5:5) is subject to the performance of the slaughtering in accordance with the Islamic laws of animal slaughter. Both of these rules are coupled together in the same context.

5. The statement of Ibn al-ʿArabī alleging that an animal that a Kitābī strangled to death is lawful is contradicted by another statement that he himself makes—in the very same book—wherein he claims that the animals of the People of the Book are lawful only if they are slaughtered in accordance with Islamic law. Of his two opposing statements, we must take the one that is in agreement with clear proof texts and the consensus of the scholars—namely, the latter. Furthermore, the incorrect statement is based on his mistaken understanding that an animal strangled to death is lawful in Christianity. We have seen from their own books, however, that this is not the case. Therefore, this statement doesn't hold any water.

6. The dominant opinion among the scholars is that it is required that the People of the Book mention Allāh's name when slaughtering, just as it is a requirement for Muslims to do so. Allāh says, "And do not eat of that over which the name of Allāh has not been mentioned" (Qur'ān 6:121). This verse applies to Muslims and the People of the Book, especially in light of the passive voice used in the words "has not been mentioned."

7. The meaning of "People of the Book" is Jews and Christians who truly believe in the basic principles of Judaism and Christianity, respectively, even if they have such false beliefs as the Trinity and Atonement. As for one who does not believe in Allāh, His messengers, or the revealed books, he is not one of the People of the Book, so the ruling pertaining to him is not like that of the People of the Book, even if such a person is statistically recorded as a Christian or Jew.

8. Meat in Muslim countries whose slaughterer is unknown is assumed to have been slaughtered according to the Islamic method, and it is permissible to eat it. However, this is not the case if it is known that the slaughterer did not do so. The proof for this is the ḥadīth of ʿĀ'isha ﷺ regarding the slaughtered meat of some Bedouin Arabs.

9. Generally, meat found in markets of the People of the Book is considered lawful unless it is confirmed that the slaughterer is not a Jew or a Christian.

10. Christians today have completely divested themselves of responsibility in the matter of animal slaughter and have abandoned the laws of their religion in this regard. They do not adhere to the legally prescribed methods of slaughter. Today, therefore, their meat is not lawful unless one knows for certain that a particular piece of meat has been slaughtered by a true Christian in a legally prescribed manner. Likewise, the meat found in Christian markets whose slaughterer is unknown is not lawful.

11. The mechanized methods of slaughtering chickens have a number of flaws in the eyes of Islamic law:

(a) Before being slaughtered, the chickens are submerged in cold water containing an electrical current. It is not ensured that the chickens do not die as a result of the electricity.
(b) It is not possible for the name of Allāh to be properly mentioned on anything slaughtered with the rotating knife.
(c) In some instances, there is doubt whether the jugular veins are actually cut.

12. We can use the mechanized system and still slaughter properly in the following way:

(a) The electrical current used to stun the animals must be eliminated, or it must be ensured that it is of an intensity low enough not to cause death before the actual slaughtering takes place.

(b) The rotating knife must be replaced with people who are stationed to slaughter and mention the name of Allāh Most High when slaughtering.

13. The mechanized method for slaughtering cows and goats has two main problems. First, it cannot be guaranteed that the methods used for stunning the animals—such as shooting a gun, using carbon dioxide gas, and administering electric shock—do not kill the animal before it is slaughtered. It is necessary to modify these stunning methods to something that is definitely not painful and does not cause death. The second problem is that the slaughtering is often not done by cutting the jugular veins. If these two problems are solved, using this mechanized method will be permissible.

14. The meat imported from non-Muslim countries is not permissible to eat, even if there is a statement on the packaging claiming that it has been slaughtered according to the Islamic method. It has been proven that these statements are not credible, and the default ruling for doubtful slaughtered meat is prohibition.

Recommendations

1. The Muslim countries should increase their supply of animals so that they do not need to import meat from non-Muslim countries.

2. If a country needs to import meat, it should try to import only from other Muslim countries.

3. Until the Muslim countries themselves reach a high level of production, their governments should mandate the importing companies to send delegates of scholars of Islamic law and experts in the meat industry to the exporting companies to demand from them modification of the current system of slaughtering to one that is in complete accordance with Islamic law. They should also appoint some Muslims who are enthusiastic about their faith to overlook the slaughtering method used by those foreign meat companies. These people should

assume a permanent role, and they should supervise the slaughtering processes faithfully. They should not issue certificates authenticating the meat's lawfulness except when they have complete certainty that a proper slaughtering took place. Furthermore, they should not issue their certificates with vague language, such as simply saying that the meat is lawful or that it has been slaughtered on the Islamic method. Rather, their certificates should be totally clear, specifically detailing that all the necessary elements of a lawful slaughtering—the animal was hand slaughtered by a Muslim or Kitābī, the name of Allāh was mentioned at the time of slaughtering, and the jugular veins were cut—have been fulfilled.

4. Until the importing companies carry out the mandates listed in (3), the Muslim governments must prohibit them from importing any meat from non-Muslim countries and from using vague phrases that claim their meat to be lawful.

5. The Islamic Fiqh Council, Jeddah, should hold a meeting and call to it as many head officials and representatives of the meat-importing companies as possible. The Council must clarify to them the significance of animal slaughter, the Islamically prescribed method for doing it, and the necessity to live up to the Council's instructions in this matter.

Appendix C

Essay on *Zabiha*-Meat Madness

Written by Muftī Abdur-Rahman ibn Yusuf
(Edited and Expanded by Amir A. Toft)

THERE IS NO doubt that there is great confusion regarding the permissibility of meats in certain countries, especially in the United States, and regarding what is lawful and what is unlawful to consume. Adding to this confusion are some of the terms used in the process.

Ḥalāl and *Zabiha*

Once while dining at a person's house in the US, the host pointed out to a guest from the UK that a particular dish on the table was *zabiha* and that another was *ḥalāl.* The bewildered guest had never heard of the term *zabiha* used in this context. Literally, *zabiha* (more accurately written, *dhabīḥa*) means "slaughtered animal." Was not all *ḥalāl* meat, the guest wondered, supposed to be slaughtered according to a particular rite, and thus qualify as *zabiha?* Why was the host then calling one dish *zabiha* and the other "only *ḥalāl*"? Could it be that *zabiha* meant something else? Maybe it was a Judaic or Christian term and this meat had been purchased from a non-Muslim source. Or did it mean super-

or ultra-*ḥalāl* meat of sorts, or perhaps signify sóme kind of organic meat? One of the local guests clarified, explaining that some people in the US use the term *zabiha* for meat slaughtered by a Muslim. This was to differentiate it from meat that many purchase from the regular markets under the assumption that it is from the People of the Book (*Ahl al-Kitāb*) and thus *ḥalāl* even though not hand-slaughtered by a Muslim.

Meat in the US and the People of the Book

Many people assume that the United States—and, for that matter, Canada, the UK, and many other European countries—are Christian nations. This is erroneous. Although many of the inhabitants of these countries profess to be Christian, this affiliation of theirs does not extend much past a personal or perhaps communal level. On the state level, these countries are openly secular, promote separation of church and state, and would hardly accept being labeled Christian nations. Some Muslim scholars from abroad, many of whom are not accurately informed of the religious composition of the US, sometimes pass the *fatwā* that the meat in the United States is lawful, citing that the country is Christian. No doubt, meat slaughtered by a Christian or Jew in accordance with the injunctions of their scriptures cannot be considered impermissible, as Allāh has made it lawful in the Qur'ān: "And the food of those who have been given the Book is permitted to you" (Qur'ān 5:5). However, carelessly purchasing or eating meat from any source in the US with the excuse of being in a Christian country and thus eating the meat of the People of the Book (*Ahl al-Kitāb*) is a gross error in judgment born out of ignorance of the true circumstances.

When purchasing meat from a regular grocery store, it is generally impossible to learn whether the slaughterer was a Christian, Jew, Buddhist, Baha'i, atheist, or, for argument's sake, Muslim; and even if it is theoretically possible to find out, the informational costs and bureaucratic barriers are too prohibitive for it to be feasible. If we then make the venturesome assumption that the slaughterer is a Christian or a Jew, the adherence of these, especially the former, to the laws of slaughter laid down in their scriptures is tenuous, as a few moments' examination of practicing Christians and Jews will show. There are, furthermore, numerous documented reports of eyewitness accounts of animals being stunned or shot and allowed to die before actually being put to the knife, thus rendering them dead before slaughtered—in a word, carrion. The US Department of Agriculture sometimes appoints representatives to watch for cadavers and remove them from the processing line, but how regular this practice is and how diligently the appointees carry out what they are commissioned to do does not inspire great confidence.

The evidence of abstaining from that which is indubitably unlawful needs no repeating. But we are faced here with a doubtful situation. In the first place, the evidence overwhelmingly suggests that the odds of finding lawful meat in US markets are slim, and probability that leans heavily in favor of one conclusion or another is recognized by Islamic law in matters of practice that bear an element of social significance, such as the food consumed by tens of thousands of Muslims. And in food-related matters, we learn from Islamic law that when the scale of surety tips toward the side of doubt, lawfulness is held in check until we can substantiate it beyond any doubt.

In general, we must be wary when confronting an issue in which uncertainty clouds the path to finding the sure ruling; we should err on the side of caution in such affairs. The Messenger of Allāh ﷺ said, "Leave what puts you in doubt in favor of what does not" (*Sunan al-Tirmidhī*), cautioning Muslims of the doubtful. From another ḥadīth, we learn a great wisdom behind this cautiousness that every Muslim ought to exhibit: "What is lawful is clear, and what is unlawful is clear. And in between the two are doubtful matters [whose rulings] many people do not know. He who guards against the doubtful safeguards his religion and honor, and he who falls into the doubtful falls into the unlawful, just as a shepherd who grazes his flock around a preserve will likely soon graze them in it. Indeed, every king has a preserve, and the preserve of Allāh are the things he has declared unlawful" (*Ṣaḥīḥ al-Bukhārī*). Therefore, abstaining from something whose permissibility cannot be established is, with little or no exception, the best path to take.

Muslims Must Demand Better

It is necessary for Muslims to bear a sense of responsibility to their faith and accountability before Allāh by demanding better of themselves and not settling with unlawful and doubtful foods. The Prophet ﷺ said, "A body that has been nourished by the unlawful will not enter Paradise" (*Shuʿab al-Īmān*). This is, therefore, a serious issue for Muslims.

Countries such as Zimbabwe, where the Muslims have been very particular about the *ḥalāl*-meat issue since they first migrated to the country, have made great progress in meat production. Beef from Zimbabwe is exported to the UK and other countries. On my visit to Zimbabwe, I was told by a member of the Council of Muslim Theologians (Jamiatul Ulama) that it had become a requirement in the Zimbabwe meat industry that all animals be slaughtered according to Islamic rites.

The concept of supply and demand should be kept in mind: if there is a demand, there will be a supply. Allāh willing, if there is a demand for an extensive range of *ḥalāl* meat products in the US and other lands where Muslims compose a minority of the population, a servant of Allāh will rise to satisfy that demand. As it has happened so far, *ḥalāl*-meat stores have had to shut down for lack of patronage from local Muslims, who prefer to purchase from their local supermarkets under the flimsy pretext that we have just discussed.

Returning to the popular terminology we introduced in the beginning, any meat that can be considered truly *ḥalāl*—that is, lawful for a Muslim to consume—is not just that it be cow meat or chicken meat or something other than pork, but that the meat be what is now in the US dubbed *zabiha,* which means, albeit inaccurately, that all the requirements of Islamic slaughter be fulfilled.

It is, furthermore, unfair, unreasonable, and childish to label the notion of *zabiha* and all the requirements it today represents as a concoction of Indo-Pakistani culture, which is the current attitude for many Muslims in the US. Granted, the fault for the misusage of this term and the consequent confusion of those like the British brother we met earlier, arguably lies with this group. But the requirements underlying it are anything but Indo-Pakistani. If we go to Syria, Turkey, Yemen, Malaysia, and West Africa, where Muslims are decidedly not Indian or Pakistani, we see these same laws upheld by Muslims who take lawful meat consumption seriously. Eating *zabiha*—or, we should say, *ḥalāl*—meat is the duty of every Muslim, period.

Our Responsibility

What, therefore, is the responsibility of Muslims living in a country like the US, where there is great confusion regarding the meat issue? What is one to do when invited to eat with friends and family who may not be very particular about their eating habits or may follow another opinion? Is one allowed to even ask of the source of their meat?

It is necessary for Muslims to ensure that what they eat is lawful. They therefore have to do their best to ascertain that the meat served in a home or bought in a store has come from a properly slaughtered animal.

Many raise the good point that excessively doubting the meat another Muslim sells or serves at home is a display of petulance that does not befit a Muslim and that inspires deep mistrust among Muslims. The author of this book himself has stated that we must think well of Muslims and give them the benefit of the doubt. But this does not mean that Muslims should cross into the territory of imprudence and naïveté. If ignorance and confusion about some matter of Islam are the prevailing conditions in a community of Muslims, ought we not tread carefully and act with a sense of caution, lest we be taken into doing something wrong by someone ill-informed? In the matter of slaughtered meat in particular, inquiry into the source of some meat is uncalled-for if we know, either with complete certainty or overwhelmingly convincing evidence, that a particular community or region or nation of Muslims follows the rules of animal slaughter laid down by Islamic law. But when the state of Muslims is such that there is sheer ignorance about or great laxity in adhering to such laws; when a community can't figure out what it means to be *ḥalāl* and what it means to be *zabiha* (or what these words even mean in themselves); and when the number of opinions about what makes meat lawful is as great as the number of those who opine; how can we so readily accept that the meat Muslims offer us is doubtlessly lawful? To be fair, this is not to attack the intentions of Muslims. Whenever we can, we must always give Muslims the benefit of the doubt in that they are well-meaning and do not *try* to serve unlawful meat to their brothers in Islam. And, admittedly, there remains yet an immense dearth of sources of Islamic knowledge, in the form of scholars or books written by scholars, to educate the Muslims about such matters (the goal of this book is, after all, to fulfill this need). Nevertheless, until Muslims become more conscious about the Islamic rulings on the weighty matter of lawful meats, we must be cautious and prepared to inquire kindly of our Muslim brothers as to where their meat comes from.

In opposition to this claim, some bring forth the following verse of the Qur'ān: "O you who believe, do not ask about things that, if revealed to you, may cause you trouble" (Qur'ān 5:101). To quote this verse in this context, however, betokens a deep misunderstanding of its intent. This verse was revealed when some companions asked questions whose answers would, at best, have done them no good and, at worse, have caused them personal grief. The commentaries of Qur'ān cite incidents where they asked personal questions bearing little importance to others, and they sometimes heavily burdened Allāh's Messenger ﷺ with a large number of such inquiries. But the source of the food we consume, in a time when few are well-informed of Islamic dietary laws, is of no little importance in the life of a Muslim. To know that food from a doubtful source is unlawful would in fact do him much good.

This verse, therefore, is not a suitable argument against inquiring into the food other Muslims present to us.

It is necessary that a Muslim do his best to ensure—by asking if necessary but without falling prey to excessive doubt—that the meat he is offered is truly *ḥalāl*, in the correct sense of the word. It is not simply enough to say *Bismillāh* on a piece of meat to make it lawful. The narration often quoted in this vein pertains, as the author expounds in the book, to a time when the meat was predominantly known to be slaughtered correctly. The Messenger of Allāh ﷺ was showing them that, in such a situation, it is not necessary to inquire into the matter, but that one should maintain a positive opinion about the one offering the meat. As for the instruction to "mention the name of Allāh and eat," it does not, as worded in the narration, conclusively imply that mentioning His name before eating makes unlawfully slaughtered meat lawful. When does a Muslim ever eat that he should not recite the name of Allāh before doing so?

On a practical level, it is sometimes difficult to determine how reliable a meat source is, and equally difficult to summon the courage and tact necessary to ask a Muslim host where his meat is from without being insulting. Here are some ways to go about overcoming these hurdles.

What to Do When Invited for a Meal

When doubtful or unsure about a person's source of meat, one can take the following steps:

1. If possible, it is usually best to go directly to the host at the time of receiving the invitation or close thereafter (but well before the date of invitation), and politely inform him or her, without being judgmental about his or her practices, that you have some strict dietary limitations regarding meat. Inform them of butchers you trust. If the meat does not come from these sources, and the host is not willing or able to obtain meat from these certified stores, let him know that meat is not

a necessary part of the meal for you and that you would be perfectly satisfied with alternate options, such as vegetable and fish dishes. Let them feel that you are eager to accept their kind invitation and enjoy their company, and that, for you at least, the importance of the gathering is in being together, not the type and variety of food served.

2. If it is difficult to approach the host, perhaps because one is not on intimate enough terms with him, then ask in an appropriate manner someone who may know where the host buys his meat. As much as possible, this should be done in an indirect and subtle manner, so as not to raise too much suspicion or throw the matter out of proportion.

3. Volunteer to purchase the meat for the host, and buy it from your trusted source. In the US in particular, many people are particular about what they eat and strictly adhere to certain self-imposed dietary restrictions, so many hosts are quite accommodating. For example, some people are on a low-carb diet, some are vegetarian, some are lactose intolerant, some allergic to nuts, and many hosts are willing to go out of their way to serve a cuisine that suits the needs of their guests.

Finding a *Ḥalāl* Meat Store or Restaurant

The matter of *ḥalāl* meat is further confounded by finding a meat market or restaurant that offers truly *ḥalāl* products. Incidents of dishonesty in some of these establishments, especially in major US cities, have most unfortunately cast doubt on whether they serve what they claim to. Some methods for finding the right store are as follows:

1. Make an earnest effort to ask around and see if some reliable and pious people in the community can provide some verification about a particular meat supplier or store or restaurant. Ask the scholars of the area, who are careful about what they eat, where they purchase their meat.

2. Speak to the store owner. Normally, they are quite open to revealing their suppliers, policies, etc. And why should they not be, if they

are not doing anything wrong? Resistance to answer these reasonable questions is a good way of gauging the honesty of a store that claims to sell lawful meat. Ask them who their suppliers are, and then ask reliable scholars in the area for verification regarding the suppliers. Unfortunately, not all store owners are forthright; some have even been found to purchase a small amount of meat from a reliable source to gain certification and then stock a large amount from unreliable or completely non-*ḥalāl* meat suppliers.

3. Sometimes there are organizations that certify meat providers. Ask reliable scholars in the area if these organizations' certifications are trustworthy.

4. Sometimes large companies that are well established and can provide their own certification in writing (for instance, that all their meat is from hand-slaughtered animals) may be trusted. However, beware of vague statements like "slaughtered according to Islamic rites," as they could be following some unreliable, unscholarly sources, or "mechanically separated chicken," which in many cases means machine-slaughtered, as opposed to hand-slaughtered, chicken.

5. If possible, team up with a few others and find a farm or abattoir willing to sell you a cow, goat, lamb, or other animal that you may slaughter yourself. Meat stores sometimes provide the cutting service of the meat for a very nominal charge.

It is not unclear that this issue is difficult to grapple with, but, as is the case with such issues, it is also a very serious one. These pieces of advice are not exhaustive and simply cannot cover the effectively infinite number of difficult scenarios one might face. The easiest method to follow is probably to ask a reliable scholar, preferably in the area, to provide some guidance for one's specific situation and, if possible, to inform you of someone trustworthy. If a person has tried their utmost in this regard, and the meat still turns out to be not what is claimed, the sin rests on the shoulders of the one who is deceitful. Allāh is forgiving but expects that His slaves put in their share of toil. As long as we are not complacent with making little or no effort, Allāh will, we always hope and pray, overlook failures that are beyond our ability to avoid.

The Past

If someone has the misfortune of having consumed unlawful meat in the past, whether because he or she simply did not know the importance of the matter or did not have a proper understanding of what meat truly qualifies as lawful, there comes to mind the agonizing question, do I really have to stop eating such meat? In a word, yes. But, like so many things, the matter is not that simple. Learning that one, having thought for a long that he has been doing nothing wrong, has in fact been doing something against Allāh's commands, can be a very shocking discovery that creates despair in the heart. It is not surprising or unreasonable that a sincere Muslim feel as if he or she can't turn back or be forgiven for having done something wrong for such a long time. But those who have this feeling should take comfort in that they are one of a large club of individuals who have this very same problem—which is to say, all of mankind. There is no human being who cannot root around in his bygone days and find some bugaboo, of whatever kind, that haunts his life, some transgression against Allāh that he feels can never be forgiven.

But anyone can take consolation and smooth out the matter by remembering the vast and incomparable mercy of Allāh, so oft-mentioned in the Qur'ān, as in the following verse: "Say, O my slaves who have committed [sins in great] excess against their souls, never despair of the mercy of Allāh! For, indeed, Allāh forgives sins, one and all. Indeed, He is the All-Forgiving, the Mercy-Giving. So turn back often to your Lord, and submit yourselves to Him" (Qur'ān 39:53–54). One must not then let Satan trick him into thinking that he has irreversibly crossed the point of no return; the only one who can shut the door of mercy on an individual is the individual himself, for so long as one seeks the forgiveness of Allāh, Allāh will be there to listen to his plea.

Those who have made this mistake in the past can rectify it and let their former errors, with the mercy of Allāh, vanish with the days that have passed. One must, therefore, repent to Allāh Most High, feel remorse over his actions, and resolve never to revert to them in the future.

And Allāh Almighty knows best.

Appendix D

"Is Islamic Slaughtering Cruel to Animals?"

Article by Dr. Aisha El-Awady
IslamOnline (Feb. 2, 2003)

ISLAM'S STANCE on what is permissible to eat and what is not is clear. There are strict rules when it comes to meat regarding what is allowed and what is forbidden. In Sūrat al-Mā'ida (The Table Spread), Allāh says, "Forbidden to you is [the eating of] carrion and blood and the flesh of swine and that over which other than [the name of] Allāh has been invoked and what has been strangled and what has been beaten to death and what has fallen to [its] death and what has been gored to death and what has been eaten by a predator—except that which you duly slaughter [before it dies]—and what has been slaughtered on any altar; and [prohibited to you] is seeking [to determine] your lot by divining arrows. This is [sheer] deviance. Today, those who disbelieve have despaired of [destroying] your religion. So do not fear them, but fear Me. Today, I have perfected for you your religion, and I have completed my grace upon you, and I am well-pleased for you with Islam as your religion. But whoever is compelled by starvation [to eat these sorts of food] without inclination to sin, then indeed Allāh is all-forgiving, mercy-giving" (Qur'ān 5:3).

The Islamic practice of slaughtering animals by means of a sharp cut to the front of the neck has frequently come under attack by some animal-rights activists as being a form of animal cruelty, the claim being that it is a painful, inhumane method of killing animals. In [many places in] the West, it is required by law to stun the animals with a shot to the head before the slaughter, supposedly to render the animal unconscious and to prevent it from reviving before it is killed, so as not to slow down the movement of the processing line. It is also used to prevent the animal from feeling pain before it dies.

German Research Studies Pain

It therefore may come as a surprise to those who have made such assertions to learn of the results of a study carried out by Professor Wilhelm Schulze and his colleague Dr. Hazim at the School of Veterinary Medicine, Hannover University, Germany. The study, entitled "Attempts to Objectify Pain and Consciousness in Conventional (captive-bolt-pistol stunning) and Ritual (*ḥalāl,* knife) Methods of Slaughtering Sheep and Calves," concludes that Islamic slaughtering is the most humane method of slaughter and that captive-bolt stunning, practiced in the West, causes severe pain to the animal.

In the study, several electrodes were surgically implanted at various points of the skull of all examined animals touching the surface of their brains. The animals were allowed several weeks to recover. Some animals were then slaughtered by a swift, deep incision on the neck made with a sharp knife, cutting the jugular veins and the carotid arteries, as well as the trachea and esophagus (i.e., the Islamic method). The other animals were stunned using a Captive Bolt Pistol (CBP). During the experiment, an electroencephalograph (EEG) and an electrocardiogram (ECG) recorded the condition of each animal's brain and heart, respectively, during the course of the slaughtering and the stunning.

Results of the Study

Islamic Method

1. The first three seconds of the Islamic slaughter, as recorded by the EEG, did not show any change from what the EEG recorded before the slaughter, thus indicating that the animal did not feel any pain during or immediately after the incision.

2. For the following three seconds, the EEG recorded a condition of deep sleep—unconsciousness. This was due to the large quantity of blood gushing out from the body.

3. After the above-mentioned six seconds, the EEG recorded a zero-level brain signal, indicating no feeling of pain at all.

4. As the brain signal dropped to zero-level, the heart was still pounding and the body rigorously convulsing (a reflex action of the spinal cord), driving the maximum amount of blood from the body, thus resulting in hygienic meat for the consumer.

CBP Stunning

1. The animals were apparently unconscious soon after stunning.

2. The EEG showed severe pain immediately after stunning.

3. The hearts of animals stunned with the CBP stopped beating early, compared to those of the animals slaughtered by the Islamic method, thereby resulting in the retention of a lot of blood in the meat. This in turn makes the meat unhygienic for the consumer.

Western-Style Slaughtering and Mad Cow Disease

Not only is this pre-slaughter method of stunning animals severely painful, as shown by the above experiment, there is also a rising concern that this method may be a factor in the spread of mad cow disease from cattle to humans. It was discovered in recent research carried out at

Texas A&M University and by Canada's Food Inspection Agency that a method called pneumatic stunning (the firing of a metal bolt into the cow's brain, followed by a pulverizing, 150-pound burst of air pressure) delivered a force so explosive that it scattered brain tissue throughout the animal. This news is disturbing, since the brain and spinal-cord tissue are the most infectious parts of an animal with mad cow disease. This stunning method causes fatal, Swiss-cheese-like holes in the brain of the infected animal. It is more disturbing to know that around thirty to forty percent of American cattle are stunned with pneumatic guns.

Islamic Regulations for Slaughter

As one can see from these studies, Islamic animal slaughter is a blessing to animals and humans alike. In order for the slaughtering to be lawful, several measures must be taken by the one performing the deed. This is to ensure the highest benefit to both the animal and the consumer.

In this regard, Prophet Muḥammad (peace be upon him) said, "Allāh calls for mercy in everything, so be merciful when you kill and when you slaughter; sharpen your blade and relieve the pain."

The object used to slaughter the animal should be sharp and used swiftly. The swift cutting of vessels in the neck disconnects the flow of blood to the nerves in the brain responsible for pain. Thus the animal does not feel pain. The movements and withering that happen to the animal after the cut is made are not due to pain, but due to the contraction and relaxation of the muscles deficient in blood. The Prophet (peace be upon him) also taught Muslims neither to sharpen the blade of the knife in front of the animal nor to slaughter an animal in front of others of its own kind.

The cut should involve the windpipe (trachea), gullet (esophagus), and the two jugular veins, but not the spinal cord. This method results in rapid gushing of blood, allowing most of the animal's blood to leave its body. If the spinal cord is cut, the nerve fibers to the heart might be damaged, leading to cardiac arrest, which results in stagnation of blood in the blood vessels. The blood must be drained completely before the head is removed. This purifies the meat, as the animal's blood acts as a medium for microorganisms. Meat also remains fresh longer as compared to other methods of slaughtering.

Therefore, accusations of animal cruelty should actually be focused on those who do not use the Islamic way of slaughtering, but prefer to use those methods that cause pain and agony to the animal and could also very well cause harm to those consuming the meat.

References

Corcoran, Leila. "Cattle Stun Gun May Heighten Madcow Risk." Reuters News Service. July 25, 1997. http://www.iol.ie/~afifi/BICNews/Health/health4.htm.

McAlister, Judith. July 12, 2002. School of Veterinary Medicine, Hannover University. E-mail: bib@tiho-hannover.de.

Naik, Zakir. "Islamic Way of Slaughtering Looks Cruel." Islamicvoice.com. September 1999. http://www.islamvoice.com/september.99/zakir.htm.

"Slaughtering and Hunting of Animals." Al-islam.org. 1997. http://www.al-islam.org/laws/hunting .html.

"Slaughter of Ruminants—Use of Pneumatic Stunning With Air Inject." The European Commission. February 17, 1998. http://europa.eu.int/comm/food/fs/sc/scv/out02_en.html.

Notes

Aisha El-Awady has a bachelor's degree in medicine as well as a master's degree in parasitology from Cairo University and is currently working as a lecturer of parasitology with the Faculty of Medicine. She may be contacted at *aawady@islam-online.net.*

Article edited by Amir A. Toft specifically for this book; editing approved by the author. Original article available at http://www.islamonline.net/English/Science/2003/02/article01.shtml.

Glossary

The following terms are listed in alphabetical order according to the English alphabet. The *ʿayn* symbol (ʿ) and the hamza symbol (') are not taken into account when alphabetizing. Macrons over the letters *a, i,* and *u* are ignored, so *ā, ī,* and *ū* have been treated equal to *a, i,* and *u,* respectively.

ʿĀDĪ. Non-ritual (literally, "normal" or "ordinary"). Opposite of *taʿabbudī.* More specifically, it refers to a matter that is not restricted by definitive rules set down by Islamic law and may be done according to how one sees fit. An example is eating an apple: no one is required to eat an apple, and there are no rules for how and at what times one must eat it; one may eat it however he pleases. Cf. *TAʿABBUDĪ.*

AHL AL-KITĀB. People of the Book, the Jews, and Christians.

ʿĀLIM (pl. *ʿulamā'*). Scholar (literally, "one who knows"). One who has been thoroughly trained and received teaching authorizations in the major Islamic sciences.

ʿALLĀMA. An honorary title used for a distinguished scholar, past or present.

ʿAQR. Method of animal slaughter whereby one who does not have physical control over the animal fatally wounds it from a distance using a sharp projectile or a hunting animal. This is mainly reserved for hunting situations. Cf. *DHABḤ* and *NAḤR.*

ARKĀN. See *RUKN.*

Aṣḥāb al-Ra'y. People of Opinion. Refers to the jurists in the early generations who placed greater emphasis than other jurists on analogical deduction in deriving Islamic rulings.

Awdāj. See *Wadaj.*

Ḍaʿīf. Weak. Status of a ḥadīth that contains some deficiency—in its chain of transmission, or one or more of its narrators, or its contents—or is not in perfect agreement with established Islamic beliefs and practices. Such a ḥadīth is generally inadmissible in deriving a juristic ruling.

Dhabḥ. Method of animal slaughter whereby one uses a sharp tool to cut the neck of the animal and make the blood flow out via the jugular veins. *See also Madhbūḥ.* Cf. *ʿAqr* and *Naḥr.*

Dhabīḥa. See Madhbūḥ.

Dhakā'. Astuteness, intelligence, mental acumen.

Dhakāt (equivalently *tadhkiya*). To slaughter an animal through the legally prescribed method that makes its meat lawful (*ḥalāl*) for a Muslim. The rules vary slightly for different types of animal slaughter, such as *dhabḥ, naḥr,* and *ʿaqr. See Dhabḥ, Naḥr,* and *ʿAqr.*

Dhakāt iḍṭirāriyya. Forced slaughter. The category of slaughtering whereby one does not have physical control of an animal and is thus "forced" to slaughter it from a distance. Primarily refers to hunting situations.

Dhakāt ikhtiyāriyya. Voluntary slaughter. The category of slaughtering whereby one does have physical control of an animal and therefore must slaughter it with the conventional, manual method.

Fatwā. Formal legal ruling issued by an authorized, competent jurist.

fiqh. Islamic jurisprudence. The science of how rulings are derived from the valid sources of Islamic law, namely, the Qur'ān, the Sunna, the consensus (*ijmāʿ*) of the Companions, and analogical deduction (*qiyās*).

Fisq. Transgression, deviance.

Ḥadīth. Literally, "a piece of news, a story, or a report relating to a past or present event." In its technical sense, it refers to the reports of the words, deeds, approvals, and disapprovals of the Messenger of Allāh ﷺ.

Ḥāfiẓ. Ḥadīth master, specifically, one who has memorized 100,000 ḥadīths by heart along with their chains of transmission. Also used for one who has memorized the entire Qur'ān.

Ḥalāl. Lawful, permissible.

Ḥanafī. Pertaining to the Ḥanafī school of Islamic law. Also, a follower of the Ḥanafī school.

Ḥanbalī. Pertaining to the Ḥanbalī school of Islamic law. Also, a follower of the Ḥanbalī school.

Ḥarām. Forbidden, prohibited, unlawful.

Hay'at Kibār al-ʿUlamā'. Board of Senior Scholars. Name of the group of leading scholars and jurists in Saudi Arabia.

Ḥijāz. Western Arabian Peninsula.

Ibn. Son. When used in a name, it usually means "son of." For example, Rāfiʿ ibn Khadīj means Rāfiʿ son of Khadīj.

ʿĪd. Holiday, day of celebration. There are two such days in Islam. One is ʿĪd al-Fiṭr, which falls after the completion of the month of Ramaḍān, and the other is ʿĪd al-Aḍḥā, which falls at the conclusion of the annual Pilgrimage (*Ḥajj*).

Ijmāʿ. Consensus. Often used to refer to the complete agreement among the Companions and also the complete agreement among the jurists of a certain period regarding a ruling for a particular juristic issue.

Ijtihād. Literally, "to struggle." In its technical sense, it means to derive the Islamic legal rulings by exhaustively analyzing the Islamic sources—taking into account their subtle implications, finer points, and wisdoms—until one feels no longer able to arrive at a more suitable conclusion. This may be done only by a Muslim scholar who has obtained an exceptional level of knowledge in Islamic jurisprudence and its prerequisite and cognate fields.

ʿIlm. Knowledge.

Imām. An honorary title for the leader of a school of Islamic law or for an eminent scholar (e.g., Imām Bukhārī). It is also generally used for the leader of the ritual prayer (*ṣalāt*).

Inhār. To cause something (e.g., blood) to flow out of its container or source.

Injīl. Appropriately translated as the "Gospel" or "Evangel." The scripture revealed to Jesus ﷺ. Over time, it was corrupted with alterations by later peoples and no longer exists with us in its original, unadulterated form. Cf. Tawrāt.

Itmām. To complete or perfect.

Jumhūr. Dominant majority. Often used to refer to the dominant majority or general body of the jurists regarding some juristic issue.

Jizya. An annual tax paid by non-Muslims under Muslim rule. Those who pay it gain the same right of protection by the government as the Muslims.

Kathīr. Much, many.

Kāfir (pl. *kuffār*). Non-Muslim, disbeliever. Anyone who rejects any of the major tenets of Islam, especially affirmation of the oneness of Allāh (*tawḥīd*) and the prophethood of the Messenger Muḥammad ﷺ. Opposite of *mu'min,* or believer.

Kitābī. Jew or a Christian, one of the *Ahl al-Kitāb,* or People of the Book. *See also Ahl al-Kitāb.*

Kuffār. See Kāfir.

Lugha. Language; linguistics.

Mā akala 'l-sabuʿu. "What has been eaten by a predator" (Qur'ān 5:3). An animal that has died because a predator attacked it and ate part of it. Such an animal is prohibited (*ḥarām*).

Madhbaḥ. Slaughtering point. The place where the upper part of the chest meets the lower part of the throat; the spot where an animal is cut when being slaughtered.

Madhbūḥ (equivalently *dhabīḥa*). An animal that has been slaughtered by *dhabḥ. See Dhabḥ.*

Madhhab. School of Islamic law.

Maḥram. A person with whom marriage is forbidden due to consanguinity (e.g., parent, sibling, child), marriage ties (e.g., mother- and father-in-law), and wet-nurse bonds (e.g., a man and woman who have been suckled by the same wet nurse).

Majhūl. Unknown. Term assigned to a ḥadīth narrator whose biographical information is partially or completely unknown.

Majūs. Specifically, the Magi, a hereditary priest class of the ancient Persians. The general meaning (and the one relevant to this book) is Zoroastrians, those who practice the ancient Persian religion of Zoroastrianism.

Makrūh. Disliked. Status of something that is undesirable or reproachable under Islamic law but not completely prohibited (*ḥarām*). Cf. *Ḥarām.*

Mālikī. Pertaining to the Mālikī school of Islamic law. Also, a follower of the Mālikī school.

MAʿLŪL. Defective narration.

MANḤŪR. An animal, especially a camel, that has been killed by the method of *naḥr*. *See NAḤR.*

MARFŪʿ. Status of a ḥadīth that is narrated of the Messenger ﷺ himself, indicated by the narrator explicitly stating "The Messenger ﷺ said such-and-such" or "The Messenger ﷺ did such-and-such" or the like. Cf. *MAWQŪF.*

MAWQŪDHA. "What has been beaten to death." An animal that has died because of a violent blow to its body. Such an animal is categorically prohibited (*ḥarām*).

MAWQŪF. Status of a ḥadīth that is not narrated of the Messenger ﷺ himself but of a companion. For example, "Ibn ʿAbbās ؓ said such-and-such." Cf. *MARFŪʿ.*

MAWṢŪL. Status of a ḥadīth whose chain of narrators is unbroken all the way to the final narrating authority (be he the Messenger ﷺ, a companion, or a follower).

MAYTA. Carrion. It is forbidden to consume.

MIʿRĀḌ. Featherless arrow. See note 19 on page 33 for detailed definition according to traditional Arabic lexicons.

MUḤADDITH (pl. *muḥaddithīn*). Ḥadīth scholar.

MUḤADDITHĪN. See MUḤADDITH.

MUNKHANIQA. "What has been strangled" (Qur'ān 5:3). An animal that has died by being asphyxiated. Such an animal is categorically prohibited (*ḥarām*).

MURSAL. Status of a ḥadīth that a follower (*tābiʿī*) narrates directly from the Messenger ﷺ without mentioning the narrator (or narrators) in between.

MUTARADDIYA. "What has fallen to its death" (Qur'ān 5:3). An animal that has died by falling from an elevated place. Such an animal is categorically prohibited (*ḥarām*).

NAHĀR. Daytime.

NAHR. River.

NAḤR. Method of slaughter that entails stabbing the animal at the bottom of its throat near the windpipe (not slitting its throat as in *dhabḥ*). *See also MANḤŪR.* Cf. *ʿAQR* and *DHABḤ.*

Naṭīḥa. "What has been gored to death" (Qur'ān 5:3). An animal that has died by being stabbed by the attached horn of another animal. Such an animal is categorically prohibited (*ḥarām*).

Qaṭʿī. Decisive. Term given to a proof text that is definitive and free of speculative content.

Rukn (pl. *arkān*). Literally, "a pillar." Technically, it refers to an essential element of some act without which the act is deficient and unacceptable.

Ṣaḥīḥ. Sound, authentic. Status of a ḥadīth that has been rigorously authenticated in its text and chain of transmission.

Salaf Ṣāliḥ. The Righteous Predecessors. The first three Muslim generations—the Companions, the Followers, and the Followers' Followers. They were the best of generations, according to the Messenger ﷺ, and they set the foundation of the subsequent generations' understanding of Islamic law and the other Islamic sciences.

Ṣalāt. Ritual Islamic prayer involving sequences of standing, bowing, prostrating, and sitting. Not to be confused with *duʿā',* which is simply a verbal supplication that a worshipper makes without going into different physical positions.

Samīṭ. An animal that has been immersed in hot water in order to remove its cutaneous protrusions, such as fur and feathers.

Shaʿā'ir. See Shaʿīra.

Shāfiʿī. Pertaining to the Shāfiʿī school of Islamic law. Also, a follower of the Shāfiʿī school.

Shaʿīra (pl. *shaʿā'ir*). Fundamental rite. One the most basic, salient, and essential features (and acts) of Islam that outwardly show that one is a Muslim and that separate a believer from a disbeliever.

Sharīʿa. Islamic law.

Sunna. Literally, "precedent" or "custom." In this book, when it is capitalized (Sunna), it refers generally to the actions and practices of the Messenger ﷺ, which together compose the second source of Islamic law. When lowercased (*sunna*), it refers to the juristic category of acts that were regularly practiced by the Messenger ﷺ and are thus strongly recommended.

Taʿabbudī. Ritual. Opposite of *ʿādī.* Refers to a matter that must be done in conformity with definitive rules set down by Islamic law and cannot be done according to how one sees fit. An example is prayer, which much be done at certain times and in a specific way; prayer not done according to the relevant Islamic laws is unacceptable. Cf. *ʿĀdī.*

Tābiʿī. Follower. Someone who saw one or more of the Companions while in state of faith (*īmān*) and died while still in that state.

Tadhkiya. See Dhakāt.

Tasmiya. Literally, "to give a name" or "to pronounce a name." Generally used to refer to the formula "In the name of Allāh, the All-Merciful, the Mercy-Giving" (*Bismillāhi 'l-Raḥmāni 'l-Raḥīm*). In the context of animal slaughter, it refers to the obligatory mentioning of Allāh's name immediately before slaughtering an animal. This statement must actually be spoken aloud, not just in the heart. And it need not necessarily be "In the name of Allāh; Allāh is the Greatest," but it must include the name of Allāh and that of no one else.

Tawḥīd. Literally, "to unite" or "to unify." It is the core Islamic concept of the absolute oneness of Allāh—the total renunciation of a father, mother, son, or any other being or thing that is above or on par with Him, and the complete acknowledgment that He is the only one worthy of any type of worship.

Tawrāt. Most appropriately translated as the "Torah." The scripture revealed to Moses ﷺ on Mount Ṭūr. Over time, it was corrupted with alterations by later peoples and no longer exists with us in its original, unadulterated form. Cf. Injīl.

Thiqa. A trustworthy authority in the transmission of ḥadīth.

ʿUlamā'. See ʿĀlim.

Umma. Literally, "community" or "nation." Used in an Islamic context to refer to the collective body of all believers.

Wadaj (d. *wadajān;* pl. *awdāj*). Jugular vein. One of the four veins in the neck (two interior and two exterior) that return blood from the head. The plural form (*awdāj*), in the context of Islamic law, refers to fours passageways in the neck: the two external jugular veins, the trachea, and the esophagus.

Wadajān. See Wadaj.

Wuḍū'. Ritual washing with water performed in order to be in a state of purity for prayer and other important acts of worship.

Bibliography

This bibliography contains a list of titles and authors of the works that the author used to prepare this book. As the translator did not have access to the precise editions that the author used, the detailed bibliographical information is not given for the majority of the works. When the complete information is given, this means that the translator did have the same edition of that particular work.

ʿAbduh, Muḥammad. *Tafsīr al-Manār.*

Abū Dāwūd, Abū Sulaymān ibn al-Ashʿath ibn Isḥāq al-Azdī al-Sijistānī. *Sunan Abī Dāwūd.*

———. *Marāsīl Abī Dāwūd.*

Abū Mujīb, Saʿdī. *Mawsūʿat al-Ijmāʿ.*

Āl Maḥmūd, ʿAbdullāh ibn Zayd, *Faṣl al-Khiṭāb fī Ibāḥat Dhabā'iḥ Ahl al-Kitāb.*

al-ʿAsqalānī, Aḥmad Ibn Ḥajar. *Fatḥ al-Bārī Sharḥ Ṣaḥīḥ al-Bukhārī.*

———. *Talkhīṣ al-Ḥabīr fī Takhrīj Aḥādīth al-Rāfiʿī al-Kabīr.*

Bukhārī, Abū ʿAbdullāh Muḥammad Ismāʿīl. *Ṣaḥīḥ al-Bukhārī.*

Danby, Herbert (trans.). The Mishnah. Oxford: Oxford University Press, 1987.

Dardīr, Aḥmad ibn Muḥammad. *Al-Sharḥ al-Ṣaghīr ʿalā Aqrab al-Masālik ilā Madhhab Mālik.*

Ḥaṣkafī, Muḥammad ʿAlāʾ al-Dīn ibn ʿAlī. *Al-Durr al-Mukhtār.*

Ibn ʿAbd al-Raḥmān, Abū ʿAbdullāh Muḥammad. *Raḥmat al-Umma fī Ikhtilāf al-Aʾimma.*

Ibn ʿĀbidīn, Muḥammad Amīn. *Radd al-Muḥtār ʿalā ʾl-Durr al-Mukhtār.*

Ibn Anas, Mālik. *Muwaṭṭāʾ Imām Mālik.* Kār Khāna Tijārat-e Kutub.

Ibn al-ʿArabī, Abū Bakr Muḥammad ibn ʿAbdillāh. *Aḥkām al-Qurʾān.*

Ibn al-Athīr, Majd al-Dīn Abū Saʿādāt Mubārak ibn Muḥammad. *Jāmiʿ al-Uṣūl fī Aḥādīth al-Rasūl.*

Ibn Ḥazm, Abū Muḥammad ʿAlī. *Al-Muḥallā.*

Ibn Kathīr, Abū al-Fidāʾ Ismāʿīl. *Tafsīr Ibn Kathīr.*

Ibn Manẓūr, Abū al-Faḍl Jamāl al-Dīn ibn Mukarram. *Lisān al-ʿArab.*

Ibn Qudāma, Muwaffaq al-Dīn Abū Muḥammad ʿAbdullāh ibn Aḥmad ibn Muḥammad. *Al-Mughnī.*

Jaṣṣaṣ, Abū Bakr Aḥmad ibn ʿAlī al-Rāzī. *Aḥkām al-Qurʾān.*

Kāsānī, ʿAlāʾ al-Dīn Abū Bakr ibn Masʿūd. *Badāʾiʿ al-Ṣanāʾiʿ fī Tartīb al-Sharāʾiʿ.*

Nasāʾī, Abū ʿAbd al-Raḥmān Aḥmad. *Sunan al-Nasāʾī.*

Nawawī, Abū Zakariyyā Yaḥyā ibn Sharaf. *Rawḍat al-Ṭalibīn wa ʿUmdat al-Muftīn.*

Niẓām, Shaykh et al. *Al-Fatāwā al-Hindiyya.*

Pānīpatī, Thanāʾullāh. *Al-Tafsīr al-Maẓharī.*

Qarāfī, Shihāb al-Dīn Abū ʾl-ʿAbbās Aḥmad ibn Abū ʾl-ʿAlāʾ Idrīs. *Al-Dhakhīra.*

al-Qurayshī, Muslim ibn al-Ḥajjāj. *Ṣaḥīḥ Muslim.*

Qurṭubī, Abū ʿAbdillāh Muḥammad ibn Aḥmad al-Anṣārī. *Tafsīr al-Qurṭubī.*

Ṣāwī, Aḥmad. *Ḥāshiyat al-Ṣāwī ʿalā ʾl-Dardīr.*

Shāfiʿī, Abū ʿAbdillāh Muḥammad ibn Idrīs. *Kitāb al-Umm.*

Suyūṭī, ʿAbd al-Raḥmān ibn al-Kamāl Jalāl al-Dīn. *Al-Durr al-Manthūr.*

ʿUthmānī, Ẓafar Aḥmad. *Iʿlāʾ al-Sunan.* 3rd ed. 22 vols. Idārat al-Qurʾān wa ʾl-ʿUlūm al-Islāmiyya, 1993.

Zaylaʿī, Jamāl al-Dīn Abū Muḥammad ʿAbdullāh ibn Yūsuf al-Ḥanafī. *Naṣb al-Rāya li Aḥādīth al-Hidāya.*

Al-Tāj wa ʾl-Iklīl bi Hāmish Mawāhib al-Jalīl.

Index

G

H

I

T

U

V

W

Y

Z

The Author

Muftī Muhammad Taqi Usmani is one of the leading Islamic scholars living today. Author of more than forty books in Urdu, Arabic, and English, he is an expert in the fields of Islamic Law, Ḥadīth, and Economics. He is the son of Muftī Muḥammad Shafīʿ Usmani, the late Grand Muftī of Pakistan. For the past forty years, he has taught at Darul Uloom, Karachi, one of the largest and most renowned schools of higher Islamic education in Pakistan. Muftī Usmani is a retired judge of the Sharīʿa Appellate Bench of the Supreme Court of Pakistan, and he currently serves as the deputy chairman of the Jeddah-based Islamic Fiqh Council of the Organization of Islamic Conference (OIC). He is also on the advisory board of a number of Islamic financial institutions around the world, and he continues to play a key role in advancing the cause of interest-free banking and making more Islamic financial institutions available to Muslims.

Muftī Usmani's works include *Takmila Fatḥ al-Mulhim,* a six-volume complement to *Fatḥ al-Mulhim,* an Arabic commentary of *Saḥīḥ Muslim* by Mawlānā Shabbīr Aḥmad Usmani. In Arabic he has also written *Buḥūth fī Qaḍāyā Fiqhiyya Muʿāṣira,* a collection of juristic discussions on some important contemporary issues (especially economics-related ones). His English works include *What is Christianity?, Islamic Months,* and *Authority of Sunnah.* The majority of his books are in Urdu, and most of them have been translated into English. Titles include *Islam and Modernism, The Legal Status of Following a Madhhab,* and *An Approach to the Quranic Sciences.* His works are widely available in the US and easily obtainable through on-line book distributors.

The Translator

Born in Chicago, Amir A. Toft completed his memorization of the Qur'an in 2002 at the Institute of Islamic Education and has studied Arabic there intermittently for a number of years. Since graduating from Northwestern University in 2005 with a BA in Economics, he is pursuing his advanced Islamic education at traditional schools in the US and abroad.

Also by

White Thread Press

Reflections of Pearls

Sufism & Good Character

The Differences of the Imams

Birth Control & Abortion in Islam

Prayers for Forgiveness: Seeking Spiritual Enlightenment through Sincere Supplication

Fiqh al-Imam: Key Proofs in Hanafi Fiqh

Imām Abū Ḥanīfa's Al-Fiqh al-Akbar with Maghnīsāwī's Commentary

Absolute Essentials of Islam

Provisions for the Seekers

The Path to Perfection

WHITE THREAD

www.whitethreadpress.com